D0077001

Endurance of Life

Endurance of Life

The Implications of Genetics for Human Life

MACFARLANE BURNET

CAMBRIDGE UNIVERSITY PRESS
Melbourne University Press
1978

Published by the Syndics of the Cambridge University Press
The Pitt Building, Trumpington Street, Cambridge CB2 1RP
Bentley House, 200 Euston Road, London NW1 2DB
32 East 57th Street, New York, NY 10022, USA

First published 1978

Printed in Australia by
Wilke and Company Limited, Clayton, Victoria 3168

ISBN 0 521 22114 5

Contents

Introduction

All men die, and in our affluent world most nowadays grow old before they die. The symptoms of old age vary almost as greatly from one person to another as their temperaments, their social worth, or the occasion of their death. In the course of my professional work on bacteria, viruses, and antibodies, I became progressively more interested in the genetic side of micro-organisms and of the body cells concerned with immunity. This led naturally to a study of how and why the immune system declines in old age, its weaknesses often allowing the terminal infection that ends the story.

In parallel, for at least forty years and for reasons personal to my own temperament, there has been a steadily firming opinion that far too little weight was being given by scholars in almost every field to the significance of genetics in human affairs. With the development of a well-worked-out hypothesis of the nature of human ageing couched in genetic terms,[1] the two streams of thought seemed to converge. This book is an attempt to bring them into a single discussion that, for me, seems highly relevant to an understanding of the darker aspects of the human situation. It touches on the three great human problems of pain, deformity, and disease; of conflict, cruelty, and evil; and of old age and death. It is the approach of a biologist, a pathologist now primarily interested in understanding the ageing process and a little excited about the only half-explored trails that diverge from the central theme.

In the past, many have written about the diversity of man or of the uniqueness of the individual, but in a world where all votes are equal, whether of citizens or of nation states, it is becoming increasingly something not to be spoken of. If a Western geneticist should aim to ape Lysenko, he would probably find a frighteningly large proportion of people in most countries ripe to seize on the heresy.

The genetic diversity of man is more relevant than any other factor to the manifestations of old age, the duration of life, and the pathology of death. Medicine in general has changed its whole pattern since in the last fifty years infection, physical injury, and malnutrition have become matters to be prevented or treated on a sound scientific basis, and effectively. What remains to be dealt with clinically is almost wholly dependent on the genetic constitution of the individual and his or her response to the social environment. Here perhaps we come to the crux of today's social and political problems: the disease-provoking and the socially harmful responses of people to the pressures of a civilization to which they are unadapted. Are those responses to be ascribed 80 per cent to inheritance and 20 per cent to the environment, or almost wholly to the damage inflicted by the injustices of society on those who would otherwise have remained healthy and conforming citizens?

In this book I want to examine those aspects of the genetic diversity of man that bear, in the first instance, on the nature of ageing and the particular hypothesis that I have adopted for the past three years. It is a genetic hypothesis concerned (a) with the replication of DNA, both in the reproductive process and during the development and maintenance of the body, and (b) with the repair of DNA after it has been damaged by a variety of physical or chemical agents. In the processes, both of replication and of repair, errors can creep in.

Error will rapidly become the central theme in my account of ageing, and from it will diverge a chapter or two on the wide range of human disease dependent on past error in DNA repair, and on the continuing generation of error and its consequences. If error in DNA is as important, as I shall show, in ageing and disease, we have to face, more critically than it has yet been looked at, the likelihood that human behaviour is predominantly, and in critical matters wholly, determined at the genetic level. Here, if anywhere, a modern philosopher will be most likely to find his approach to the problem of evil in the world. We may find in the end that war and evil, pain and disease, ageing and death were inevitable as soon as a working pattern of life with DNA, ATP, and protein as enzyme had been devised. If this is in any acceptable sense the truth about the human situation, it must have many implications in most fields of human thought and action. To those who can see some merit in this point of view, it will almost certainly be regarded as a pessimistic approach that has something of the same demeaning effect on human dignity that Darwin's ideas had 120 years ago. This may be true, but it is no argument against the validity of the approach.

I am very much aware of the gaps in my knowledge of the relevant fields that I have drawn on, but, as I shall emphasize repeatedly, the complexity of biology is such that no human being can have a comprehensive knowledge of more than one very limited area. Similarly, I know from past experience that I shall be accused of a complete lack of subtlety in handling matters of human thought and behaviour. Yet subtlety in current writing is almost wholly dependent on producing sequences of words that in any scientific sense are meaningless but which for those of sensitivity and literary background are designed to offer an appreciation of thoughts and emotions that are clearly too complex and incommunicable to be expressed in language intelligible to a biologist in his role as scientist. Good literary English is impossible without an imprecision that would be intolerable in good scientific writing. I have deliberately tried to write on topics that lie just outside the accepted realm of science in the same way that I would undertake an article on immunology for *Scientific American* or some equivalent medium.

At the age of 78, I have no axes left to grind, and in trying to shape aspirations for the short term or for a million years I have done so to the best of my ability in terms that would in principle allow, in due course, a scientifically valid assessment of the outcome. I have tried to extract the relevance for human affairs of modern work on genetics, on the nature of error in biology, and on the relative importance of inheritance and environment in determining differences among people's mental characteristics and behaviour. In doing so, I am sure to have been ignorant of many things and shown lack of judgement about others, but I do not admit that those lapses have been due either to self-interest or to fear. In essence, what I am seeking is an understanding that might directly or indirectly help toward something better than the twisted and tortured history of man for the last ten thousand years.

1

Origins of Life and Death

It may be sheer arrogance for a gerontologist to claim that to understand his subject one must begin at the very beginnings of life. Yet to all of us there comes a time when the experience of old age plus the imminence of death become of overriding personal importance. For most people the comfort of traditional religion, the local patterns of ritual, and the affection of kinsfolk are enough to make that phase as bearable as other circumstances will allow. It may be that the intellectual, the scholar, the scientist has no justification for bringing his special skills to bear on natural death and what precedes it. In the handling of old age, the doctor, the priest, and the welfare officer have a part; a greater part is asked from family affection and care, but little from the philosopher.

But since the first flowering of what became European civilization in classical Greece, there have always been some people with a greater capacity to grasp the meaning of things, and a much larger number who admired and rewarded that capacity. In every century the world has had to be looked at again. In fact, when in any field there appears a new concept, a new alignment of power, or even a new technique, a freshet of intellectual endeavour is set in motion. Scholars and scientists are always at hand, eager to exploit anything that is new, and the world is always ready to applaud any success that they achieve. If medicine and biology are shedding new light on the nature of ageing, that in itself is a justification for looking at the human aspects of old age and death from the new angle. Perhaps the most important human problem is raised by the disappearance of any support, philosophical or scientific, for a continuation of personal consciousness after death. There is a virtual taboo on public discussion of that conclusion, although it is probably accepted by 99 per cent of those with a reasonable understanding of physiology and by a very substantial proportion of intelligent men and women. If the expectation of immortality has vanished, some other means of allowing other forms of satisfaction with life in old age seem to be called for.

My title, and to a considerable extent the whole theme of the book, derives from two and a half well-known lines from Shakespeare's *King Lear*:

> ... Men must endure
> Their going hence even as their coming hither.
> Ripeness is all. Come on.

The words are spoken by Edgar to his old and blinded father, Gloucester, as he leads him away from the final downfall of Lear. It is said in reply to Gloucester's hopeless comment that a man could rot even here. Old age, injustice, pain, disability, and hopelessness may be inescapable, but one must go on to the end.

One might almost define the objective of the welfare state as to rid its citizens of the fear of those things which Gloucester had no alternative but to endure. To define how far that objective is possible and desirable, and to plan how it may be achieved, will require the exercise of hard, well-informed intelligence, as well as compassion and political expediency. Part of what will be needed is an effective understanding of man's origins as a living organism, a mammal, and a human being. This and the following chapter are intended as a reminder, a recapitulation, of the story of evolution on earth insofar as it appears to be relevant to the theme of human old age and death.

Life's origin

Life as we know it is a phenomenon that was generated on our own planet. This is the almost universal conclusion of those with adequate understanding of the evidence. The mode by which solar systems develop as stars of the main series take shape is well understood. One of the consequences is that it now seems certain that there are vast numbers of solar systems broadly comparable to our own in the universe. Many of them will have planets at that optimum distance from the sun that should allow life to develop as it has done on earth. Movement of life from one solar system to another can be ruled out. Time, distance, and the lethality of exposure to interstellar space leave, as the only possibility, transfer in some space vehicle far more sophisticated than anything so far even conceived on this planet.

The process by which life arose on earth has been discussed many times since the first writings on the subject by Oparin and Haldane early in this century. It is clear enough that life must have evolved in liquid water and the process could not have begun until the first seas were well established on the cooling earth. The situation then must have been such that an almost infinite variety of relatively small organic molecules could form at random in the warm, primitive seas. Between that stage and the appearance of self-reproducing micro-organisms equivalent to primitive bacteria there is an enormous conceptual gap. But the time gap is also large, possibly as much as 1000 million years. If that primitive organism was created by an accumulation of random events—*and there is no other conceivable process within the universe known to science*—the chance of its occurring must have been almost infinitely small. But it did occur, and all that can be said is that, if the circumstances are right—the numbers of organic molecules in the upper layers of all the water on earth, the infinite variety of accidental occurrences, and the almost infinite lapse of time represented by 1000 million years—anything that can happen will happen.

Somehow the micro-organisms that have been found in South African rocks 3000 million years old came into being. They must have had a chemical structure based on DNA and protein and a turnover of protein and energy handled by enzymes. Once the organismal pattern had been laid down, there is no special intellectual difficulty in accepting that an evolutionary process could develop the general sequence of organisms that we can follow through the geological record. Most of the really difficult and interesting things must have taken place in that first 1000 million years. Without making any attempt to speculate about the chemistry of what took place, one can consider some phases that could be relevant to the understanding of death.

The primitive meaning of death

Death is clearly impossible until we have some entity that corresponds to our definition of life or living matter. Probably the most basic definition is that life is the capacity to reconstruct a replica of a complex pattern of organic molecules. This would include some laboratory artifacts, but that does not disqualify it. As soon as large molecules could build up larger complexes by bringing in smaller molecules, and these complexes break up into two or some other small number of units with similar capacity, a process *toward* life was under way. That process would at first be fragile in the extreme. An abrupt rise in temperature or a switch of the water to a more acid or a more alkaline reaction would probably destroy the system and in a sense kill the tentative approach to life. True death could hardly be spoken of until micro-organisms had evolved as self-replicating units with a compact structure, controlled by DNA, and building up its substance prior to division into daughter organisms, by taking in small molecules and abstracting energy from the environment in one way or another.

Death now comes to mean any event that disrupts the biochemical processes of the unit in such a way as to make any continuation of replication permanently impossible. In a sense those first primitive bacteria, like their modern equivalents, were potentially immortal. A bacterium enlarges and divides into two; apart from accident or error, both are precisely identical, both are of the same generation. One may even guess that the bacteria one finds in any scrap of dirt represent the last proliferation in an immortal line that stretched back to the inconceivably ancient bacteria of the primitive seas. This holds equally for any other contemporary organism, including ourselves. For someone still with a capacity for wonder, it can be fascinating to look through a microscope at one of his own white blood cells. He can see the round central nucleus and he knows that in the nucleus are strands of DNA. It is the literal truth that the patterned molecules in that DNA have come down in an immortal, unbroken sequence for 3000 million years from the single micro-organism within which the universal genetic code first took its definitive form. Life in that sense is immortal, and in the early stages, when reproduction was no more than growth and division into two, there was no immediate biological necessity for death. But multiplication will always have to stop when food supply is exhausted. Among the bacteria, as for every other organism, survival will on the average leave only the same number of individuals alive as were alive a day ago, a year ago, or a century ago, in a sufficiently similar environment. All the other myriads that were produced will be dead. Death can come in an infinite number of ways, and in most species, whether animal, plant, or micro-organism, the selection of individuals for death is intensely random. Every organism is an intricate mechanism and it can only continue to function effectively and survive if its structure is not significantly altered from that laid down by its genes. How gross any damage by trauma, heat, or infection must be to be significant, i.e., lethal, will depend on the particular construction of the organism involved. A spade plunged into the soil will push thousands of micro-organisms aside without harm, will cut an earthworm in half, leaving both halves to survive and regenerate, but kill almost at once if it chances to sever a field mouse. In a mammal, as we know only too well, any major disturbance that allows blood to flow unchecked, that severs a vital nerve tract, or that introduces a chemical poison to disturb the body's regulatory mechanisms will be lethal.

Death may be essentially at random, but chance has its own regularities. Whenever bacteria are growing in any natural accumulation of water and nutrients, changes in the proportion of the different bacterial types are constantly taking place. To reduce that situation to a laboratory model, let us think of a fluid containing two nutrients, X and Y. We have at hand a mixed culture of two sorts of bacteria growing in a fluid that contains only X and we transfer a million or two to the X + Y mixture. What happens will depend on whether any of those bacteria can grow at the expense of nutrient Y as well as X. If one of the strains can do this and the other cannot, it will obviously flourish. Both will multiply and both sorts will lose many by death, but the one that can use Y will be at an advantage, and after a time it will probably be found that the population has become composed *only* of bacteria that can use Y. If there are large enough numbers in the populations concerned, any new option, however minor, will favour the survival of those organisms that can exploit it.

Predator and prey

As evolution moved into the period of the conventional fossil record, which began with the Cambrian Period 600 million years ago, predation had become the major cause of death. From the very beginning a compact micro-organism was a potential source of nourishment for any larger type that might evolve means to break it down

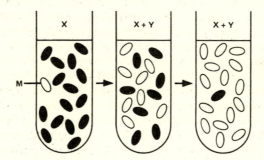

Fig. 1:   *Adaptation in bacteria. In the original culture containing only nutrient X, there are a few of another type which may have arisen by mutation or in some other way. If this minority (shown as white) can, unlike the original (black) form, utilize nutrient Y as well as X, successive transfer of the culture in fluid containing X and Y will result in rapid replacement of the original (black) population by the new form.*

into fragments capable of being utilized. The first animals came into being as eaters of bacteria and microscopic algae; and soon afterwards no doubt bacteria appeared that found they could enter and exploit the substance of some of those primitive animals as a source of food, thereby introducing infectious diseases into the world. Predator and prey, parasite and host—the two sets of relationships became established almost as soon as life took on organized form. There is little in the story of life that is not intimately based on one or the other—and not least in human history.

Life has a strange and frightening difference from the other phenomena of the universe, from the other shapes that matter has taken. Once the possibility of reproduction had been achieved, it became inescapable that, having regard to the 'state of the art' at the time, life must always expand to convert the largest possible amount of the earth's substance into living material. Every organism has always produced many more offspring than could survive, and once the life process was under way most of those that did not survive were destroyed and their substance made use of by other living forms. The green plants, using the light of the sun as their source of energy, provided the basic food store for all. They were the only innocents, and even from the green plants there arose eventually the insect-eating sundews, a number of parasites on other plants, and the strangler fig trees of tropical rain forests.

Evolution in essence makes use of the genetic plasticity of living things to devise new ways or more effective old ways for an organism to prosper, to occupy a new ecological niche, as the phrase goes, and to increase the total amount of living material, the bio-mass, in the region. The limiting factor, once the green plant had developed, was solar energy, and the primitive earth soon developed a green cover over the whole of the land surface where there was adequate water and a warm enough climate for trees or grasses to grow. It was an efficient process, and the food synthesized by the plants was eaten by animals and broken down by micro-organisms. The plant eaters in their turn provided food for the carnivores, and an approximate balance developed. Populations would wax and wane, but over decades the counts of different plant and animal species would remain approximately constant. In the seas the green plants at the basis of all the food chains were the microscopic algae. The food chains themselves were longer and more complex than on the land, but predators and parasites played essentially the same roles.

Two aspects of these ecological qualities of the earth 200-300 million years ago need some comment in relation to the theme of ageing and death. The first is that the use of solar energy to produce

organic material that could be used for food almost regularly produced more than could be eaten. Fire from lightning got rid of
much of it, but where forests grew on marshy areas much of the
organic material degenerated to peat, lignite, and coal. The sources
of petroleum are still problematical, but in all probability it too
represents organic material greatly changed but having escaped
oxidation by biological processes or by fire. So the fossil fuels
accumulated to wait for the twentieth century.

Ecology, which is concerned with the interactions of organisms,
plant, animal, or microbial, with each other and the environment,
provides the background to evolution. In dealing with birds or
mammals, it is concerned with how individuals of the species make a
living, the population level at which the species is in balance with the
environment, and what changes in population are going to result
from change in this or that relevant facet of the environment. Evolution works by changes within the species population as a whole, or,
when we look at it, as we should, from the genetic angle, it is by
change within the open gene pools of the species.

Gene pools and DNA

Gene pool is so important a concept that it may be worth using a few
paragraphs to consider it, and before we can start doing that a word
must be said about DNA—deoxyribonucleic acid—the key molecule
for the whole of life.[1] Probably almost every reader knows at least
that DNA makes up the long, complex molecules that, using what is
called the genetic code, contain the information needed to synthesize
and control the substance and function of the cell. We exist because
all the information to produce us was there in the DNA of the
fertilized ova from which we derive.

In the very real sense we can say that the continuing thread of life
is DNA; the animal or the plant is, as it were, a mere mortal
excrescence hanging on the network of potentially immortal DNA.
The network is the gene pool, and, as DNA can exist only in a living
cell, its material basis is the specialized reproductive cells of each
individual. An open gene pool is the sum of the genetic information
in all reproductively capable individuals who can freely interbreed.
In the last analysis it determines the distribution of characteristics
and potentialities of the next generation of the species. Later on,
much more will have to be said about the intensely random character of evolution, but at this point the first requirement is to introduce
the concepts of phenotype and genotype. Without use of
these two terms, discussion of genetic influences in evolution is
almost impossible. Any organism, insect, mouse, or man, develops

its characteristic structure on a basis of the information carried in the DNA and present in the nucleus of every cell of its body. That mass of information constitutes the genotype of the organism; it will differ to some extent from that of any other individual of its own species, and of course much more from that of any other species. Differences in the genotype amongst individuals provide the raw material from which new combinations of qualities can be brought together to allow the workings of evolution.

An animal's genotype is something that must be deduced by genetic reasoning from its visible and scientifically demonstrable characters and those of its progenitors and descendants. Only part of the information in the genotype is manifested in the organism, and what is expressed is intensely dependent not only on the intrinsic genetic processes but also on environmental factors at every phase of development. In the embryonic stages everything depends on the internal environment which is based in part on maternal contributions as well as on the progress of the embryo's own development. At later stages the influence of the outer environment in all its complexity becomes equally important. Two organisms of essentially the same genotype may differ widely in their size and other demonstrable qualities as a result. The form and function that a mature organism presents to the world is for the biologist its phenotype. It is on differences of phenotype from one individual to another within a species that natural selection works.

Natural selection is a term to cover all those circumstances that are responsible for the universal observation that, of the organisms born, only a proportion, and often only a small one, will survive to reproduce. Essentially accidental factors will obviously play a part in determining which individuals will survive, but chance is not everything. To take an imaginary example, we can think of a population of 10 000 field mice whose main predator is a species of owl. In that population we can separate out a subpopulation of 10 per cent who are more adept than the others at escaping from owls. At the end of the year we might find that 90 per cent of the standard form had been eaten, but only 70 per cent of the more agile group. The ratio between the two, instead of 90 per cent to 10 per cent, has become 900 to 300, i.e., 75 per cent to 25 per cent, and if the process of multiplication and survival went on for a few more years it is obvious that soon there would only be mice of the more agile sort left. This is a simple example of a stochastic situation where basically random conditions applied to a large enough population of animals or any other sort of units generate a recognizable regularity that can be expressed in numbers.

Some elementary genetics

To go further in understanding the changes in the gene pool by random predation and accident as modified by genetic variation in the probability of survival, it is necessary to move more directly into genetic concepts. Genetics is concerned primarily with the process by which the information in the maternal reproductive cells is, as it were, split in half, one half being discarded and the other half being carried and supported in the ovum. A process quite similar in principle results in the sperm also carrying a half complement of chromosomes and information. With fertilization, the full number of chromosomes (the diploid number) is reconstituted. What is important to grasp is the part chance plays in these processes, and for this the analogy of the two packs of playing cards can be useful.

We can think of an organism that has a diploid number of eight chromosomes, four pairs, in each of which one comes from the male, one from the female, parent. We can take each suit as representing a chromosome and we start off with two complete packs separated into four double suits. In the two sets of spades the two aces will correspond but they will be recognizably different in the way the ace is depicted. There will be similar minor differences between the two deuces, and in general for every other pair of cards in the pack. The first procedure, corresponding more or less to the reduction division, is to shuffle the two spade suits and then go through them, putting the first of each pair to appear on the left hand side and the other on the right, when in due course it emerges. Again, we have two full suits of spades, but by a process technically called crossing-over they are no longer the same as the original male- and female-derived chromosomes. The actual process of crossing-over is not quite of this type, but for the present purpose the shuffle is sufficiently close in principle. Eventually, we have the four suits in sets of two. From each we take one set at random and discard the other. Those we retain represent the single set of chromosomes (the haploid number) for ovum or sperm. Every ovum will end up with a somewhat different combination of genes from any other, and the same will hold for the sperms. At fertilization, one or a small number of ova will be fertilized very much at random by the winners of a race in which there are some hundreds of millions of competing spermatozoa. In terms of our packs of cards, the set of four suits carried by the ovum is now matched with the corresponding suits carried by the successful sperm, giving the new diploid combination of two sets of each suit which is different from that of either parent and also from that which would have resulted from fertilization by any other sperm.

All the genes of the new zygote (the fertilized ovum) come from the parents, but for each offspring they are combined in a different way: siblings are notoriously different one from the other, but, even so, it is usually possible to detect contributions from both parents in the appearance and personality of each. Any reader who has an intimate association with a large family will be able to confirm this and in doing so will gain an important insight to the most important source of human diversity, this universal process of sexual recombination of genes. For the present it is unimportant where and how the original differences arose. They have been present throughout mammalian evolution, and in the way I have described they have been redistributed at every fertile mating.

Another aspect of sexual reproduction is more directly relevant to the evolution of ageing. When every mated female of a species produces each year many (from 10 to 10 000) offspring, and the expectation is that around two will survive to reproduce in their turn, ageing is a meaningless word. Consider what happens to the 'family' of 10 000 ova produced by a fish that completes its life cycle in a year or two. In broad terms it is something like this. A very high mortality among the unhatched ova and the earliest young, due to all manner of accidents and predators, will probably reduce the population to, say, 10 per cent by the end of the first month. Thereafter, steady predation will proceed to deplete the population in approximately exponential fashion. With the passage of each month, an almost constant percentage of those present at the beginning will have been eaten or otherwise eliminated by the end of the month. As the survivors increase in size and in experience, the rate of predation will diminish, but, even so, by the end of the year there will probably be less that ten remaining. For most marine animals such a sketch may not be far from the reality. Populations are dynamically balanced to contain a more or less constant number of mature individuals. In any relatively stable ecosystem there is a constant turnover, with death always at hand to ensure that nothing will block the continuing evolution of the species.

Death as accident

But life has all sorts of manifestations, and most generalizations will have exceptions. The largest of the bivalved molluscs are the giant clams (*Tridacna*) of the coral reefs in the Pacific and Indian oceans. These have massive shells that can reach a size of a metre or more in their longest dimension. They appear to be immune from predators and have developed a co-operative (symbiotic) relationship with algae, which provides important elements of their nutrition. No one can say how old the largest examples of the giant clam may be, nor is it known whether they die of old age. What seems most likely is that as the shells become more and more massive they encroach on the internal space available and diminish the efficiency of the organism in various ways, perhaps involving the all-important reproductive capacities. To use the giant clam as an example of a general evolutionary phenomenon in the absence of any real knowledge of the facts is obviously foolhardy, but it will probably make a theoretical point clearly enough. Evolution is concerned only with what happens before reproduction ceases, and in all probability its concern diminishes rapidly after the first phase of mature reproduction. By that time any mutants or recombinants that might have an advantage for survival will have had their chance. If some adaption has its optimum survival value in early maturity and for essentially geometrical reasons continued growth beyond that time hinders reproductive efficiency, it is vanishingly unlikely that evolution will do anything to right this disability. This is a form of a generalization about ageing due to Medawar. He held that if some unfavourable effect of a certain combination of genes was constantly occurring, one of the ways evolution could, as it were, forget about it was by pushing its manifestation progressively into the period of old age.

Another important possibility is that death may come to any long-lived organism by simple accident, i.e., by some damaging episode that is too rare to have any significance as an evolutionary factor. For the giant clam, one can imagine that gross damage to its coral environment by an exceptionally severe cyclone, an abnormal appearance of toxic micro-organisms in the environment, and probably many other rare circumstances could provide a chance of lethal damage.

Nature has many ways of handling the necessity for death. Among invertebrate forms there is an immense variety of life styles and lifespands range widely. The duration and quality of any phase of senescence is known for only a few species and is probably relevant in each case only to closely allied forms. With a central commitment to human ageing and death, however, any biological models and analogies must be drawn from the vertebrates, and almost exclusively from mammals. I shall therefore conclude this chapter with a few notes on ageing in the lower vertebrates and leave a more comprehensive account of the findings in mammals to Chapter 4.

The most primitive vertebrates are the cyclostomes, with two distinct families now in existence, the hagfishes and the lampreys. In the fossil record the first vertebrates to appear were the ostracoderms, which were probably heavily armored cyclostomes. Immunologists have found much of interest to study in the lamprey, but I do not know of any attempt to measure lifespan or to gain any other information about changes with age in the immune system or elsewhere.

Among the bony fishes there are some tropical forms that have a lifespan limited to one year. Walford has studied one of these 'annual fish' in detail, with the interesting finding that lifespan was considerably prolonged when the fish were kept in subnormal temperatures and the characteristic ageing changes in collagen (see page 73) slowed down. Another example of interest among fish is the Atlantic salmon, which, once spawning in small fresh-water streams has been completed, undergoes remarkably rapid anatomical changes and dies within two or three weeks. Little is on record in regard to amphibia or reptiles, although these of course include the notoriously long-lived giant tortoises of oceanic islands. Overall, what is known of ageing in cold-blooded vertebrates suggests that throughout the lower vertebrates is has hardly been necessary for evolution to consider the need for an ageing process except where some aspect of life style, as in the annual fish or the Atlantic salmon, calls for some special development. Where pressure for survival is minimal, as amongst the island tortoises, growth and life seem to go on indefinitely. There are some dominant species, too, which are known to be long-lived and among which some particularly large specimens are occasionally found. Big estuarine crocodiles may be up to 6·7 metres (22 feet) long, and one is recorded as 10·06 metres. Both European and American sturgeons can reach very large sizes, and examples seventy to eighty years old are on record. The simplest interpretation, though not necessarily a universal one, is that the norm for evolutionary survival is the standard sexually mature form and that thereafter there is no selective modification possible although growth continues as before. Under these circumstances, growth would inevitably reach a point where sheer unwieldiness or some other physiological handicap from excessive size would predispose to death. Equally relevant is that sooner or later in an indefinitely continuing life a lethal accident must occur. In other words there may be no genetic processes that can define a period of senescence in such forms. Always with the reservation in mind that if some evolutionary need for a defined lifespan arose from the particular life style of a species, means to ensure it could be devised, we can probably think of the lower vertebrates as not being genetically programmed for senescence. All those that survive as species have been able to reach a tolerable balance between births and deaths that allows predation to be the most important factor in preventing overpopulation.

When this general view is adopted, death becomes quite unimportant. It is a simple consequence, to be worked out under the influence of time and chance, of the various factors under genetic control which are there to ensure that the species should survive. All that is needed is that on the average they give rise to an adequate population in each generation, large enough to carry the potentialities of variation that will let the species take advantage of circumstances that might allow better use of a changing environment. If that is achieved, death becomes a random, inevitable event, almost irrelevant to biological realities.

However much this may hold for most invertebrates, it is not necessarily so for man and the other mammals. The main current task of gerontology is to ascertain for mammals how far the sequence of senescence and natural death has a meaningful genetic basis that transcends mere accident.

2

Complexity and Error:
the Emergence of DNA

Man—body and mind—is the most interesting and most complex aggregation of atoms in the universe as we know it. Yet the human organism has an evolutionary history that can be followed back with some confidence, despite the gaps, to the very beginning of things. In some way the formless flux of elementary particles and energy that made up the primordial fireball was capable in due course of generating all the complexity and diversity of the world today. The fact that, implicit in the 'Big Bang'—that utterly improbable occurrence with which the universe began—was the potentiality of the emergence of man and of intelligence, provides the least answerable of all philosophic questions. I am not competent to discuss the philosophical side, but I can follow in outline the steps by which the atoms of the elements appeared and were distributed in their observed frequency through the cosmos. The origin of the solar system is relatively well understood as the gravitational concentration and redistribution of an immense cloud of cosmic material. Each planet took its final form largely in accord with the distance of its orbit from the sun. The earth in the early stage of its formation lost most of the lighter and more volatile substances, and, as it cooled and consolidated, a second redistribution of its matter liberated a primitive atmosphere and the water that became the oceans and gave origin to life.

The significance of error

In Chapter 1, I discussed some aspects of the first stages of life in the upper levels of the seas; here the approach is to the significance of error. In a sense, the whole road to man has been a catalogue of errors. Most of the matter in the universe is still in the form of hydrogen, and only an infinitesimal part of the hydrogen is in the form of water molecules. In turn, only a small proportion of water is liquid. Life as we know it—and most likely *any* organization of matter that could be spoken of as living—depends on the presence of water, liquid water, which can exist only over one small range of temperature. Terrestrial life is only possible from 0°C. to about 60°C. In the universe as a whole we have a range from absolute zero (-273°C.) to a temperature around $100\,000\,000$°C. in the centre of the stars. Life could never have emerged except on a planet where there was an abundance of water and a surface temperature that kept most of it liquid most of the time. The occasional appearance of a planet at just the right distance from its parent star must be an accident affecting only a tiny proportion of the matter in the pre-stellar cloud, an error in the orderly sequence of stellar evolution. It is almost axiomatic that in any system significant *change* can involve only a small part of it. This becomes far more evident when the system is a living one.

In Chapter 1 an outline was attempted of how life arose on the earth by some very rare accident among the enormously extensive and varied population of molecular aggregates present throughout the surface waters of the primordial seas and oceans. In considering what followed, it is necessary to bear in mind the time-scale concerned (see Figure 2). The formation of the earth-moon system, with approximately the same atomic constitution as now, took place about 4500 million years B.P. (before the present). The earliest

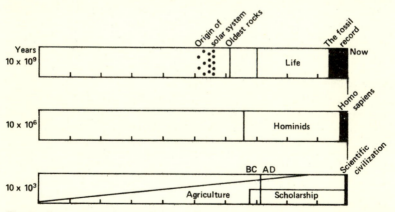

Fig. 2:   *Terrestrial time: with three scales, in units of one thousand million (10^9), one million (10^6), and one thousand (10^3) years. The 'fossil record' spans 600 million years. Modern man may be 250,000 years old. Scientific technological civilization can be dated from the Great London Exhibition, 1851, or soon afterwards.*

sedimentary rocks (about 3400 million years B.P.) show that water was already present as liquid seas and oceans, and that cycles of erosion and sedimentation were well established. The earliest microfossils with a reasonable—but not certain—claim to represent micro-organisms are dated around 3000 million years B.P. From about 2500 million to 2300 million years B.P. stromatolites appear in the geological record, which from comparison with similar modern structures are interpreted as mineral precipitation on sheets of primitive blue-green algae, the first chlorophyll-containing micro-organisms. The proterozoic era is therefore dated from 2500 million to 600 million years B.P. This includes the Precambrian, where in specially suitable rocks impressions of soft-bodied invertebrate forms have been found. The standard beginning of the fossil record of animals and plants is at the beginning of the Cambrian Period, 600 million years B.P.

Happenings at the biochemical level in the period before 2500 million years B.P. that led up to the first appearance of self-reproducing units would be of immense interest to chemists, and there are many speculations. From our biological point of view, however, such details are irrelevant and we can start with a micro-organism essentially equivalent to a modern bacterium, at least as far as its mode of reproduction is concerned. This would necessarily imply many other similarities to the modern forms, and anything to be said will have to be based on an elementary understanding of the genetics of some modern prototype bacterium. By a combination of accident and design, the standard form used in virtually all laboratories that are interested in bacterial chemistry and genetics has come to be the common intestinal organism *Escherichia coli*, known to every biologist as '*E. coli*'.

Three thousand million years ago the surface waters must have swarmed with such primitive micro-organisms living without oxygen, presumably on organic molecules in solution including fragments of dead and disintegrated micro-organisms. In the absence of oxygen in the atmosphere, much lethal ultraviolet radiation from the sun must have penetrated to a moderate depth of water and killed a relatively large proportion of the organisms near the surface. In the very, very distant future, descendants of these micro-organisms would evolve into algae and other chlorophyll-containing plants that in due course would produce the molecular oxygen of the atmosphere and allow the development of an ozone shield against most of the destructive effect of the ultraviolet. In the meantime, however, one of the main tasks of evolution must have been to develop ways by which DNA could be protected from ultraviolet light and by which any limited damage produced by ultraviolet could be repaired. The evolution of the enzymes responsible for replication and repair of DNA must have taken place at this period and established the pattern that exists in today's human cells as well as those of every other animal. Only when such matters had been dealt with could it become possible to develop the standard processes by which evolution has worked over at least the last 600 million years. Before that, all is conjecture, but in a real sense evolution has been continuous since the first atoms larger than hydrogen emerged from the primordial plasma. It is only the nature of the *significant* errors that has changed and continues to change as complexity and organization increase.

DNA and its replication

Before going on in logical fashion to discuss the genetic structure of the simplest organisms we know today, it is necessary to remember that our primary concern is man. DNA is DNA, whether in the simplest bacterium or in our own cells, but the simpler form is of course much easier to study in the laboratory. Only by studies on *E. coli* was it possible to combine chemistry and genetics to give us an understanding of the genetic code and the structure of DNA as a 'double helix'. A bacterium, however, has no more that a primitive and rarely used process of sexual interaction between the DNA from two individuals. The immense importance of sexual reproduction for the evolution of higher forms is only dimly foreshadowed in the bacteria. For any reader to whom DNA is a much used but only vaguely understood combination of letters, it seems important to introduce here something about human DNA and a little more elementary genetics.

Every child receives his inheritance in almost equal parts from his mother and his father. When an egg cell has been fertilized by a sperm and is ready to begin to develop, it contains two sets of genetic

Table 1

DNA (4)	A (adenine)	C (cytosine)	G (guanine)	T (thymine)
RNA (4)	A ———	C ———	G ———	U (uracil)
PROTEIN (20)	ALA	ARG	ASN	ASP	CYS
	GLN	GLU	GLY	HIS	ILE
	LEU	LYS	MET	PHE	PRO
	SER	THR	TRY	TYR	VAL

Table 1: *The three alphabets.*

information, one from each parent, in the set of coiled coils of DNA and protein that we call chromosomes. The DNA, the famed double helix, is what actually carries the information which in due course will be expressed in the structure and activities of the embryo and then the adult animal. It is near enough to the truth to think of DNA as an almost endless tape on which there are two rows of symbols, each like a string of letters in a line of typescript. But there are only four letters in this alphabet—A, C, G, and T (Table 1). These four symbols actually represent rather large organic molecules, called nucleotides, made up of a base, a sugar and phosphate, the letters being the initials of adenine, cytosine, guanine, and thymine, the bases that give individuality to the nucleotides. In genetics we are concerned, at least at the elementary level, only with the symbols. The order of the letters conveys the information just as it does in printed words, but there is one additional point to be remembered. Our tape is a rough diagram, as it were, of the double helix, in which two single strands of DNA twist spirally around each other and are held together by bonds between the units we call A, C, G, and T. The rule as to which bonds with which is very simple: it is A with T, and C with G. In Table 2, a short length of double-stranded DNA is shown to indicate how the sequence of nucleotide units conforms to this rule. When DNA duplicates itself so that two daughter cells can each receive precisely the same information, the helix unwinds over short lengths and each of the separated strands picks up a new set of partner units according to the same A-T, G-C rules. This of course will give both the 'daughter' double strands exactly the same structure.

With that in mind, we can revert to the simpler models that we believe are more or less equivalent to primitive bacteria of 2000 to

Table 2

DNA (A)	..	A	G	C	T	T	A	G	G	C	T	..
(B)	..	T	C	G	A	.A	T	C	C	G	A	..
RNA	..		UCG		AAU		CCG		A			..
PROTEIN				— SER — ASN — PRO —							

Table 2:   *The relationships of DNA, messenger RNA, and protein. Part of the DNA of a structural gene with two complementary strands. On transcription of the upper strand, the corresponding length of RNA is shown in triplets (UCG, etc.). These are translated into structural amino acids of the primary polypeptide of a protein.*

3000 million years ago, *E. coli* and one of its viruses, the bacteriophage T4. Everything suggests that what can be learnt from these is essentially applicable to all higher forms, subject only to the greater complexity and greater control that are needed when organisms with many other functions not represented in the bacteria come on the scene. What came before the typical bacterial DNA of today may eventually be elucidated, but nothing useful can be said now for those interested in general biology. One can take *E. coli* and T4 together because most of the significant virus-bacterium interactions concern their DNAs and in their particular host-parasite association it is essential that the DNAs should be mutually compatible. Without attempting to elaborate the statement, one can say without fear of contradiction that there has been a cross transfer of bits of DNA between host and parasite over the ages. So much so that it would probably never be possible to sort out those stretches of nucleotides that are 'really' of bacterial origin and those that are proper to virus.

Bacterial DNA is in the form of an immensely slender thread, a double helix that forms a complete circle with about 1 000 000 nucleotides per strand. Those two strands contain equivalent amounts of information and when they are untwined section by section each can provide a template on which a complementary strand can be constructed in the process of DNA replication.

The most fundamental activity of *E. coli* is its proliferation when supplied with a suitable nutrient solution. All young biologists are introduced to the idea of exponential growth by tales of how a single organism whose descendants were kept under optimum growth conditions would produce about a kilogram of bacterial substance in twenty-four hours and a million million tonnes in forty-eight hours. Fortunately all exponential processes must slow down and stop before an impossible situation is reached. The essence of life is its capacity to increase the number of individuals at a rate which will rapidly fill any suitable environment, any new ecological niche, with great rapidity. Generation time increases with increasing size; a virus will produce 100 new virus units in the time it takes one *E. coli* to become two, i.e., about half an hour; a fruit fly will produce a new adult brood in less than two weeks; a pair of mice will produce a litter by the time they are nine weeks old. Even man, living in conditions of optimum opportunity, can show a population doubling time of thirty years or less. Compared with geological times of the order of tens of millions of years, any living organism multiplying without restraint can overrun the earth in a twinkling.

To return to *E. coli*, the most important aspects of its proliferation is the duplication of DNA, which is the necessary beginning of the process of growth and division. In recent years it has grown more and more obvious that the replication of DNA is an extremely compliex process, and far from being just the automatic pairing up of the proper nucleotides that we tended to picture when the Watson–Crick double helix first broke on the world.

As far as one who has never worked directly in the field can gather from recent reviews, the first step is a series of unwindings of short segments (Okasaki pieces) of DNA in which an 'unwinding protein' is involved. At the end of each segment an incision is made and a short sequence of RNA inserted as an initiator for the replication process. This requires an elaborate set of enzymes co-ordinated in some fashion to find the right place and insert the right number and sequence of ribonucleotides. Another elaborate process follows by which the new strand is gradually lengthened, nucleotide by nucleotide, each being matched against the complementary nucleotides, A with T, G with C, on the template strand. When one tries to picture the process in detail, one can realize the need at each step for a number of distinct enzymes to recognize from the template what nucleotide is called for, to extract the proper one—A,G,C, or T—from the pool, which must constantly be replenished from where the nucleotides are themselves being synthesized. As the segment is completed, other enzymes come into action to remove the temporary RNA primer, rejoin the segment to the completed part of the new DNA, and ensure that the two strands twist around each other to form a correct double helix. Indirect evidence suggests that some of these enzymes are constantly travelling around the circlet of DNA to recognize and repair minor damage, particularly a loss of the purine bases characteristic of A and G nucleotides. The fact that these are spoken of as editing or monitoring enzymes is itself an indication of how elaborately controlled the DNA system is, even in *E. coli*. The total number of nucleotides in the double-stranded circular chromosome is a few hundred thousand. When the bacteria are multiplying freely, the whole sequence of nucleotides is serving as a template for the production of an exactly similar circle, the process of DNA duplication being completed well within the thirty minutes required for cell duplication.

I have emphasized both the complexity and the high degree of co-ordination needed for replication and also the fantastic speed and accuracy with which it is accomplished. Admittedly this is the key process in every organism and a mechanism which has been subjected even in *E. coli* to 3000 million years of evolutionary refinement. Other basic functions of the human body, such as the structure of the neurone and the distribution of synaptic junctions or the molecular dynamics of mitochondria, those powerhouses of enzymes in every cell, seem on incomplete evidence to be at least as sophisticated functionally and at the molecular level. The point I am making is the extreme complexity that is immediately revealed when we approach as has been possible only within the last two decades—the molecular realities of biological structure and function. The minute size of cells and cellular mechanisms is highly relevant in any consideration of whether any effective understanding of function within a human or any other type of cell will ever be attained and, even more so, when attempts at highly sophisticated types of manipulation of cell mechanisms are contemplated, e.g. 'genetic engineering', to remedy inherited defects in infants.

Only in bacteria and viruses do we have fairly detailed information about genetic function. A standard human cell, such as one of the fibroblasts from a piece of skin that we can persuade to grow in culture, has some 800 times as much DNA as there is in *E. coli* and at least an equivalent degree of related complexities. In particular, each human (mammalian) cell has its own specialized and limited function in contrast to the uniformity, independence, and all-inclusiveness of bacterial cells. Mammalian cells are vastly more elaborately organized than bacterial cells, even though their degree of specialization means the absence of some capacities present in *E. coli*.

This is enough for the present to express the immense and necessary complexity of every fragment of living matter. Every cell is equally a fully co-ordinated, meaningful working system with no shadow of random quality. Almost as in the days of Paley's *Evidences*, every organism, every cell, can be regarded as a miracle of design. Some attempt, necessarily partial and amateurish in quality, is needed to describe or, rather, hint at how effectively functioning design can arise from the random process that was briefly outlined earlier. This has been done much better than I can hope to achieve by Jacques Monod in *Chance and Necessity*,[1] but he was concerned almost wholly with bacteria and I am for the most part dealing with man.

Mutation

For good reasons I shall take the process by which mutations arise after exposure to ultraviolet as a prototype of organic change, and to introduce such a discussion something must be said about the nature of protein and of enzymes. A small protein is made up of about 400 amino acid units strung together like a string of beads (a polypeptide chain, so called) and coiled or crumpled into some definable secondary configuration. Those amino acids are set in their correct order by a controlling sequence of three times that number of nucleotide A, G, T, C units, which makes up the DNA of what is called a structural gene. Throughout the body we find structural protein giving form and strength to the bones and muscles and making up much of each tissue and organ. Intimately associated with structural protein in the vital organic substance of the body is catalytic enzyme protein responsible for the synthesis and interactions amongst all the substances, all the molecules, of the body. It is the task of an enzyme to bind two simple sugars together to make lactose, but it is equally the responsiblity for at least a hundred enzymes to co-operate in the production of a protein molecule and ensure that it conforms strictly to the information in the DNA sequence of a structural gene. The whole functioning of the body depends on the multitude of enzymes being at the right place at the right time and in the right concentration; above all, it demands that each enzyme is produced strictly to its pattern as laid down in the gene. Error in the structure of a gene will be reflected by error in the composition of the enzyme whose synthesis it controls, error which is liable to render its catalytic action unduly weak or absent and sometimes to produce secondary error in the product of the enzyme's action. Error is likely to be particularly dangerous if it involves an enzyme whose action is crucial in ensuring accurate production of any of the 'information-carrying' large molecules, especially DNA. In such a case, secondary errors could occur to involve *any* structural gene being synthesized in the cell. It is easy to see that once such a process of error producing further error is initiated, it must sooner or later result in complete disorganization of the function of the cell.

In addition to structural genes there is a variable amount of DNA in the chromosomes which appears to exercise control functions, the most obvious of which is to specify the time when a given structural gene will be turned on (derepressed) or turned off (repressed). There is every reason to believe that control or regulatory DNA has essentially the same structure as structural DNA and is replicated and repaired by the same enzymes; in other words, it can suffer error in very similar fashion. Unfortunately, little is known about either the normal function or how normal function can be modified by informational error in regulatory DNA, and at a later stage I shall follow precedent by using what is known about structural gene function as a basis for speculation about control DNA.

DNA damage and repair

This is perhaps sufficient background to allow a return to our central immediate problem of the nature of mutation following exposure of genetic material to ultraviolet radiation. The general quality of mutation has been touched on already at several points, but it is crucial enough to justify an attempt to make the process as clear as is possible without doing complete injustice to the complexity of the situation and the gaps in accepted knowledge.

When ultraviolet light impinges on superficial cells like those of the skin or on cells experimentally exposed in thin layers of fluid, a standard type of damage is inflicted on the DNA. Where there are two adjacent T nucleotides in one or other of the two strands, they are liable to be fused together by the occurrence of a chemical reaction between the T bases which characterize the T nucleotide unit. Technically this is referred to as thymine dimerization, but its importance for the present discussion is that the existence of such damage makes it impossible to duplicate the segment of DNA involved. As a result the cell is unable to multiply and, in any experimental investigation, if dimers persist in the DNA the cell is registered as dead.

There are two ways of repairing that fusion of bases. The first is the most direct. By providing a special enzyme to do the job, *E. coli* inactivated by ultraviolet can recover to a large extent if soon after the inactivation it is briefly exposed to visible light. The light provides a quantum of energy which the active enzyme can channel

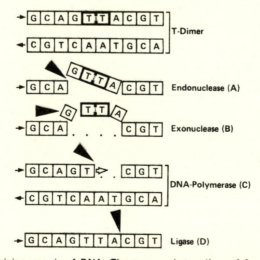

Fig. 3: *Excision repair of DNA. The successive action of four enzymes in removing the T dimer produced by ultraviolet light and correctly repairing the gap. The point of action of the enzymes is shown by the dark wedges.*

into producing a direct reversal of the chemical union. The general view is that this process of reactivation by light is rather a special attribute of bacteria and does not occur in human cells. There are other more complex but also more versatile repair processes which are demonstrable in all types of cells, from bacteria to human skin cells. Identical or very similar processes are also effective against other types of DNA damage, such as that produced by X-rays or by various chemical mutagens. It will serve our purpose to describe only the simplest of these repair processes since they are all variants on that theme.

The process can be outlined by saying that enzyme A seeks out and recognizes the damaged spot; then two or three units before and after the point of damage it makes two nicks in the chain of nucleotides making up the damaged strand. Enzyme B then sets to work at the first nick and removes all the nucleotide units up to the second nick. Enzyme C is the only one we need name; it is a DNA polymerase (DP for short) whose role is to replace the missing nucleotides correctly. It is a very complex task and, as might be expected, it needs a complex enzyme or, better, an enzyme complex to do the job. DP may in fact be a complex of ten or twelve different proteins and it is probably always a little prone to error even though it includes, as I mentioned earlier, monitoring, editing enzymes in the complex. Finally, enzyme D is swung into action to tie in the new piece and make good the integrity of the strand.

Informational error

In this sketch of the process I have asked for some extraordinary biochemical feats to be done at a molecular level and with only a fraction of a second for each unit of action. It is on record that Charles Darwin, when he looked at the peacock's tail, was moved almost to doubt whether evolution could have produced it. A biochemical geneticist may be similarly driven to wonder how DNA could possibly arrange its own repair in the way I have outlined. But the peacock's tail in display and the repair of DNA in a sunburnt skin are both facts of life in the most literal sense. Once again, one can only believe that there has been world enough and time for even the most improbable events to have occurred repeatedly. Usually the miracles of life are triumphantly achieved, but there is always the possibility of error. Sometimes the miracle does not quite come off.

And in saying that, we are beginning to approach the great revealing paradox of life that it is only the errors that have made the miracles possible.

Let us return to the cell in which perhaps a hundred points of damage to DNA have been repaired, some by the process I have superficially outlined, some by an even more elaborate procedure proportionately more liable to error. Damage to DNA is not a mutation, though it can be the main initiating factor of the process that results in a mutation. A rather tortuous definition of a mutation is to say that in the process of repair such as I have just described, and the rather similar sequence in the course of duplicating normal DNA, error can occur by which the sequence of nucleotide units in a strand of DNA is distorted; an A unit may be replaced by G, half a dozen units may be lost, or a small random group inserted by error. Once anything of the sort happens, new activities come into play that with proper evolutionary reservations one can say were 'designed' to get round the difficulties these abnormalities present. For anyone interested, there is plenty of discussion, much of it speculative, about various expedients that are or may be available. It is enough to say that in one way or another a solution is often reached by which an error can be accepted and the cell regain a capacity to grow and divide like a normal cell in every respect except for that change in the nucleotide sequence. There is an *informational error* in the patterned sequence of nucleotides that controls the cell and it must have its effects on the cell's function.

To picture the sort of effect an informational error in DNA can have, one might no doubt obtain some glittering analogies from modern techniques in the manufacture of miniaturized computers, but to attempt that is beyond my competence. As an old-fashioned writer who provides handwritten material for his typist to transcribe and punctuate, I am more familiar with errors of transcription. In the draft typescript, even the most expert typist will make an occasional error. The great majority are easily recognizable and make no difficulty at all in understanding the meaning of the sentence. Sometimes, however, a little word easily mistaken for another will lead to trouble; its accidental replacement may still give a legitimate sentence but with a wrong meaning. With my hand writing, 'on' and 'or' look very much alike, as do 'none' and 'more'; with either pair a change can be disastrous.

The analogy with DNA and its errors is rather close. Most small informational errors will have only a trivial effect but, very rarely, an error in one unit of a gene will be reflected in a single change of an amino acid amongst the 500 or more in a protein and produce something as disastrous as sickle-cell disease.

In presenting this rather simple picture of mutation as genetic error in replication, I am very much aware of the complexities that are waiting around the corner with each significant advance. In *E.*

coli there is already evidence of a logical complication which will certainly have its analogies in human cells. When the bacteria are heavily irradiated and much complex repair work has to be done, it seems *as if* a new set of DPs was called into action to handle the exceptional problem, a set which is more versatile but less faithful in its replacement than the normal enzyme complex. There are other alternatives and one of them may be that when the more complex type of repair is needed, the nature of the repair enzyme complex is such that it must *produce* more errors.[2]

We have come then to the point where we have a working idea of the way in which an informational error may be implanted in DNA and persist through all its subsequent duplications. That error may alter the structure of some protein and modify its functional efficiency to a greater or smaller extent. It may equally modify the processes by which regulatory genes control development and maintenance. In all probability, most of the qualities we recognize as anatomical mutations—tumours, pigment collections, misshapen features, etc.—are due to mutations affecting control DNA. And there is a further point of the utmost importance. If the structure of the enzymes that handle DNA repair is itself changed by mutation, it is almost inevitable that the *rate* of mutation in cell or organism will be altered. In most cases any change from the normal will mean that the complex is more error-prone and the rate of mutation will be increased.

Recombination

Mutation is the raw material of evolution, but in man and the higher organisms it plays virtually no part in evolution as such. Errors and mutations are always accumulating over the generations, but the great majority are of little or no significance. Nearly all are pigeon-holed, as it were, by being stored as 'recessives'. In general, when a mutation involves a certain gene, the equivalent gene, its allele, is normal and its normal information overrides the faulty information from the mutated gene. There are exceptions but they do not seriously infringe the general rule that most mutations in organisms more complex than bacteria have no immediate effect. They become important when two affected alleles of the same gene come together, one from each parent. If the abnormality in both has a similar effect on bodily function, this will be expressed in the offspring. Usually, the effect will be harmful or neutral, but on rare occasions it will be beneficial, particularly if the relevant environment has been changing. Once the effect of the gene has become manifest—or, as we say, been expressed in the phenotype—it becomes subject to the test for survival and to a wide variety of chance vicissitudes—included in the concept of genetic drift—which can sometimes even allow a mildly disadvantageous gene to prosper.

Experience and logic applied to human populations and to wild populations of fruit fly, *Drosophila*, indicate that in both species almost every individual carries a proportion of recessive genes which if they were phenotypically expressed in double (homozygous) form would give rise to some demonstrable and disadvantageous change of structure or function. In man, the number has been estimated to average somewhere between five and ten per individual. In addition to these skeletons in the genetic cupboard we can be certain that there are much larger numbers of neutral or nearly neutral mutations also awaiting opportunity for recombination. It is from such new combination that most of the advantages that we recognize as contributing to evolutionary progress arise.

To many people this accumulation of tiny improvements by randomly accumulated mutations seems completely inadequate to account for the eagle's eye, the peacock's tail, or the human brain. The only answer is to stress that the whole process of sexual reproduction ensures a maximal likelihood of testing out every possible mixture of the latent potentialities of past mutations yet at the same time retaining most of them in the gene pool for further testing. Perhaps one should add that the sequence of mutation and recombination, selection and survival, as the basis of evolution had been clearly formulated as a result of the work of Darwin, Weismann, Mendel, Haldane, Morgan, Sewall Wright and Fisher by 1940. Over thirty-seven more years it has been found adequate and there is probably no biologist of standing who would not support the dictum that there is just no other conceivable interpretation of the history and nature of life.

In summary then, life long ago discovered in DNA an immensely versatile mechanism for, one might say, converting error into information. The possibility of erroneous insertion of nucleotide units in the linear strands of DNA must have been present since the first DNA took form. As the enzymes controlling DNA became better organized, error-proneness diminished but never became zero and could be changed when required, e.g. in relation to the repair of different types of DNA damage by mutagens. Errors accumulated more or less constantly during evolution and were either lost by selective processes or in due course were incorporated into standard pattern.

At any one time the genetic pool of a species contains large numbers of mutant alternative configurations constantly being tested and retested to find any new combination that will provide a new phenotype with survival advantage. Especially when human genetic change in the relatively short term is under consideration, the store of existent mutations in the gene pool is sufficient to provide large numbers of new combinations of qualities and to allow a redistribution of qualities within the pool if environmental circumstances change sufficiently. Mutations will continue to occur, but in the context of historical time limited to a few centuries they are likely to have only minimal effect. Only in genetically and geologically significant periods of the order of millions of years are the mutations going to play a dominant role.

Unless there is some undetected weakness in the logic of biological science, one must conclude that the evolution of man has conformed to that of all other organisms. His present structure, capacities, and potentialities for achievement, like all other contemporary manifestations of life, have come into existence on the basis of chance occurrences—by error or accident. Mutations are essentially random occurrences, a proportion of which are rendered meaningful by natural selection; which, in Julian Huxley's words,[3] 'achieves its results by giving probability to otherwise highly improbable combinations—and in the teeth of a storm of adverse mutations.'

Whether what we recognize as the progressive emergence of ever more complexity and novelty of quality through the whole evolutionary process includes something inexplicable on any current hypothesis can probably never be excluded. The most evident difficulty is to understand how a brain-mind, evolved simply for the better survival of primitive man and his hominid ancestors, could on occasion reach the heights of achievement that have immortalized Rembrandt, Newton, Beethoven, or Einstein. Yet no alternative appears even conceivable. I prefer to leave the problem as unanswerable and try to accept the natural limitations of the machine we think with.

3

Development of a Personal Approach

Once one has retired from a scientific career and is interested in writing in the main because it is the best way to clarify one's own ideas, there is no special reason for staying firmly within the fields one has cultivated at a professional level. As an immunologist I had been well aware how the efficiency of the immune system diminishes in old age, certainly in man and mouse and probably also in other mammals. For many years that was only a peripheral interest, and it is perhaps not wholly a coincidence that the first lecture I gave on the specific topic of ageing was in 1970, and I am just three months older than the twentieth century. Maybe I was beginning to feel that old age was upon me. Nevertheless, there had been much incidental interest in aspects of age in relation to human disease and speculations that somatic mutation might be related to the ageing process, for at least twenty years before that. Like many other intellectually curious physicians, I had spent many hours playing with the mathematical regularity of the causes of death in relation to age.

Statistical information on which to work is easily obtainable in any large medical or general library, and for the major countries it has been available for many years. The state collects much information about its citizens, but the most consistently demanded and most accurately recorded are the day on which a person is born and the day on which he dies. Tabulations will also be found stating the 'cause' of each death that was shown on the death certificate. There is much scope for error in regard to cause of death, and anyone using the data must be circumspect about his conclusions. Some causes of death, such as cancer of the breast, cerebral haemorrhage, pneumonia, or physical violence will be correctly stated in well over 90 per cent. On the other hand, there will be many individuals who die after a chronic illness and are reported as heart failure or bronchopneumonia, in whom the most important factor in the final outcome is either something unsuspected and not looked for or some complex of circumstances that even an experienced pathologist with full post mortem information cannot fully interpret. However, with proper circumspection it is possible to find many interesting regularities in the age incidence of mortality from most of the important common diseases.

My interest in the age incidence of disease was a catholic one, first stimulated by the interesting differences between the common infections and with special reference to poliomyelitis and influenza, which were current subjects of laboratory investigation. Apart from incidental discussion of the age incidence of these two infections, my only publication in the field dealt with the age incidence of infectious disease in infancy and childhood. However, at various times I have played in different ways with the regularities in statistics of cancer and respiratory infections, with benefit I hope to my sense of the significance of 'random regularities' but without publications.

There were hints in both those groups of diseases that immunity, fading with old age, had some influence in the steadily rising incidence of such deaths with each additional year of life. The meaning of this inescapable weakening of the immune system with age gradually became one of the minor but still fairly popular fields for immunological research. In 1970 I was much impressed with the possibility that ageing might be wholly a result of the running down of an immunological clock and that perhaps the thymus was that clock. I began to read more widely and became more and more convinced that though immunity might be important it could not be the whole story. But I had no special interest in making ageing a central object of theoretical study.

Radiation Dangers

That interest arose in a curious fashion in the early months of 1973. A new Labor government had just come into power in Australia, and among its election promises was a firm stand against pollution of the environment. The nuclear bomb tests being carried out by the French government in the South Pacific represented a politically valuable target, and among other activities the Prime Minister asked the Australian Academy of Science to appoint a committee of experts to assess the dangers to human and animal life in Australia of these French tests in the atmosphere. I had been for two years (in 1957 and 1958) chairman of the Commonwealth Radiation Advisory Committee, and probably for that reason became a member of the 1973 committee.

The Australian government wanted an early reply, and our first discussions made it clear that all we could usefully do was to apply to the local situation the internationally accepted rules. Fortunately, two excellent discussions of the problem and codification of rules and recommendations had recently been published, one by a United Nations committee and one by the United States authorities. Also available were good records of atmospheric radiation levels over Australia, from around 1954 onwards. Over the years since the French tests started, the records clearly showed the effect of the plume of radio-isotopes rising from the French explosions and carried from west to east round the world to Australia. One could rather inaccurately work out the total additional amount of radiation that people in Australia would be exposed to as a result. The international committees had provided a set of practical rules for action based on their assessment of the danger that very small amounts of radiation represent to human populations. Our task was to apply the rules to predict how many people would suffer from leukaemia and thyroid cancer and how many genetic abnormalities could be expected in the Australian population. We did the sums to the best of our ability and made our report. None of us was very happy with the outcome, and in the report there is a paragraph to point out that the figures were obtained by taking the 'worst possible' set of circumstances. We could not exclude the possibility that the effect might be very much smaller.

I found that committee period of extraordinary interest at the personal level. I 'did my homework' conscientiously, particularly in regard to what is called the dose-effect relationship, and very soon came to the conclusion that on current knowledge there was no realistic way of assessing whether the minute amounts of radiation with which we were concerned were doing any harm or not. Our sums merely assessed the worst possible result if certain rules that were regarded as reasonable were used in the calculations. The whole procedure could almost be equated with the mediaeval discussions as to how many angels could dance on the point of a needle. Reading between the lines, one could see that the overseas authorities had much the same view but felt that where matters of public health and safety were concerned the 'worst possible' result must be taken as the criterion for preventive or protective measures.

What interested me most was the recent development of an understanding of the way in which DNA—the key substance responsible for cellular inheritance—could be effectively repaired after it had been damaged by atomic radiation or ultraviolet light. This had not been allowd for in those 'worst possible' estimates, thereby in my opinion destroying most of their usefulness. DNA repair was unknown in 1957-8, when I had my earlier official associations with radiation dangers, and I now read all I could find about it. In the process I had also to become reasonably up to date in regard to what was known about mutation.

Mutation research first became important when Muller in 1927 found that fruit flies exposed to X-rays developed increased numbers of the rare inheritable abnormalities that appear spontaneously in plants and animals and that had been known as mutations for at least twenty years previously. By 1973 many chemical agents were known that were highly effective in producing mutations, and few people doubted that such chemical mutagens acted very similarly to and were perhaps identical with the cancer-producing substances we call chemical carcinogens.

My interest in radiation, mutation, and DNA damage and repair lasted long after the committee had sent in its report, been thanked by the Prime Minister, and disbanded. All through, one particular facet had interested me: that the spontaneous appearance of cancers and leukaemias or of genetic abnormalities in human populations was very much greater, sometimes a thousand times greater, than could be accounted for by any mutagenic radiation present in the environment. Nor did it seem even conceivable that the result could be due to unsuspected chemical mutagens taken in with air, water, or food. Evidently mutations affecting the whole body—germ-line mutations arising in the reproductive cells and giving rise to inherited abnormalities—usually arose spontaneously. This held too for the leukaemias that for good practical reasons are used as an index of somatic mutation, i.e., mutation occurring in a body cell other than those cells that are set aside for reproduction. I felt that I should like to know what really caused these 'spontaneous' mutations. Out of this there arose my current professional interest in ageing and the eventual genesis of this book.

Somatic mutation

I have already mentioned the term 'somatic mutation' in contrast to germ-line mutation. The phrase was first used by Tyzzer in 1916 in relation to cancer, and since then one can say broadly that cancer theories have alternated between those that ascribed cancer to virus action and those that in one form or another blamed somatic mutation. In 1957 I rather firmly committed myself to support of the somatic mutational side, and have never found cause to change that viewpoint. Indeed, my interest in somatic mutation has steadily extended into other areas of the medical sciences rather faster in fact than is acceptable to the majority of people who are interested in such matters. My views on immunity, and particularly on auto-immune disease, were largely based on somatic mutation, and in 1965 I was talking of ageing being largely a result of somatic mutation, though perhaps mainly mediated through auto-immune processes.

Freckles

At this point I must obviously attempt to explain as clearly as possible just what is a somatic mutation. It is not easy to find a really suitable example, and many geneticists might accuse me of over-stepping the bounds a little by taking the freckles of childhood as a prototype of somatic mutation. To the best of my knowledge, man (and Caucasian man at that) is the only animal who freckles when exposed to sunlight. Freckling is normally regarded as one of the nice things that happen to children. It is not a disease and no one is going to spend years of research on the nature of freckles, particularly as to do it properly would involve removing representative freckles surgically at frequent intervals. Most of what is known about freckles comes from simple and rather casual observation. There is, however, a serious genetic disease where the most conspicuous symptom is dense freckling over exposed areas of skin. The disease is known as xeroderma pigmentosum; it is very rare, but has been so successfully studied in the last decade[1] that I have devoted Chapter 9 to its discussion and feel fully justified in using the freckle as an example of somatic mutation, even though in doing so I must make a number of statements from analogy with other conditions.

A freckle is an accumulation of pigment cells just beneath the outermost layer of the skin. Commonly it is a circular disc formed by a single layer of cells and showing a uniform brown coloration. Some children have almost uniform freckles varying only in diameter, but in the lightly pigmented 'thin' skin of a red-headed small boy it is usual to see a fine and varied crop of freckles. They are of various sizes and often show up to four or five grades of pigmentation. Each freckle starts as a just visible point and expands into a disc which, having reached a certain diameter, tends to stay constant. Their eventual fate seems to be to fade and be lost in the gradual darkening of the skin with age—but I have found no account of the process by anyone who has followed it systematically.

So much for the natural history of freckles, and now an attempt to reconstruct at the cellular level how they arise. Once again I must stress that this account is derived in part from analogy and imagination simply because the necessary experimental studies have never been made.

In the normal skin there is a layer of pigment cells at the base of the epidermis (the outermost layer of skin cells). These cells are quite different from the epidermal cells. They are more closely related to nerve cells, and actually their progenitors migrate during embryonic life from a structure in the developing nervous system called the neural crest, to the skin. The pigment cells, melanocytes, distribute themselves evenly in their particular stratum in numbers that are related to the degree of skin pigmentation. The first hominids, the first creatures to be recognizably man-like and from whom we are descended, arose in Africa and were black. European skin pallor is a later characteristic, and the best reason I have seen given for its evolutionary development is an interesting one: to provide more of vitamin D, the substance that prevents the bone disease rickets. The vitamin is made in the skin, under the action of ultraviolet light, from precursor substances produced in the body. An inadequate supply is crippling, and without vitamin D survival would be impossible. Deep pigmentation of the skin minimizes damage by the ultraviolet fraction of the sunlight, but it also diminishes the capacity to synthesize the vitamin. The spread of human beings into cold, cloudy countries could only succeed if skin pigmentation diminished as they infiltrated northward over hundreds of thousands of years. Europeans are not properly shielded against sunlight, and when they move to sunnier climates they are liable to suffer unduly from sunburn, skin cancer, and, when they are young, freckles.

The melanocytes are labile cells capable of multiplication in response to stimulation, as by sunburn, but as long as they are genetically normal they can distribute themselves evenly so that skin colour is uniform over considerable areas and changes only gradually according to the degree of exposure to light. In a freckle we are dealing with altered mutant cells that arise as a result of the mutation-producing (mutagenic) action of ultraviolet light. The thin epidermis of a child blocks only a little of the light falling on it, and ultraviolet can be damaging to skin cells, both melanocytes and the basal epithelial cells from which the epidermis develops. The most important part of the damage is to DNA. Here and there in cell nuclei a strand of DNA is chemically altered by the light so that it cannot divide to produce another similar strand. This means that the cell itself cannot divide and is effectively killed—unless the DNA damage can be repaired. Fortunately it usually *is* repaired and in nearly every instance the repair is properly done, bringing the cell back to normal in all respects. Sometimes, however, there is an error in the repair. The sequence of chemical units in the repaired DNA strand is not the same as it was in the original, and when the cell starts to function, some inadequacy or abnormality appears. The cellular abnormality reflects in a way that I have already described the error in the DNA strand, and since that altered strand pattern is repeated with every duplication of DNA, the cellular abnormality also appears in every descendant cell. The change in the cell is referred to as a mutation—a somatic mutation—and the cells of the new line (or clone) as mutant cells.

Mutations occur in random fashion and the type of change can vary enormously. Most mutations are never recognized. A single cell is far below the limit of ordinary vision; we should need great magnification before we could see the skin as being closely peppered with pigment cells. In fact if we think for a moment, it will be clear that almost the only way we could recognize a mutation in a skin melanocyte is if the mutation results in *proliferation* of the cell to produce a more crowded collection of brown cells big enough to be seen.

Mutation of melanocytes can become visible as freckles, as larger pigmented areas that dermatologists call café-au-lait spots, as moles, or as the cancerous development of a mole, which is the very dangerous malignant melanoma. They are brown spots, patches, or lumps; in all of them an error has occurred in the processes that control proliferation and distribution of melanocytes. These accumulations of cells are necessarily derived by descent from a single mutant cell. They constitute a *clone* of genetically uniform cells. As I have already noted, there are many other types of mutation that do not produce clonal proliferation, and in view of our primary interest in ageing, it is worth while to think about some of these inconspicuous mutations.

If some minor fault develops in a piece of machinery, the usual result is that in one way or another the machine is less efficient than it was. This holds equally for mutations. The normal cell is something that has been perfected over millions, in some aspects over billions, of years. Any random change is vastly more likely to reduce efficiency than to produce some desirable modification. A casual genetic error in a melanocyte is likely to reduce its capacity to produce pigment, or to distribute itself evenly in the skin. I have only to look at the back of my hands to see, in addition to some little rough white patches that come from mutation in epidermal cells, a few pale patches, as well as one or two larger irregularly shaped darker patches of pigmentation. These changes reflect more than seventy years' exposure to light, and doubtless many thousands or millions of somatic mutations. In addition, there is thinning and wrinkling of the skin itself, due no doubt to loss of damaged cells that could not be effectively repaired. The most likely story is that, as random errors accumulate in a cell, the most common terminal event is death of the cell. Atrophy is a general manifestation in ageing tissue. There is some evidence too that skin pigmentation decreases in old age, perhaps from a progressive loss of melanocytes.

Ageing and cancer

From this example of skin pigment cells and their changes one can derive a general idea of the nature of somatic mutation and how it may be of importance in relation to the process of ageing. As I shall discuss in detail at a later stage, this is now a widely accepted general basis for ageing. The attitude, however, is relatively recent, and I can complete this chapter by continuing a personal account of my own development as what might be called an amateur gerontologist. Perhaps what was most influential was my special interest over the last decade in autoimmune disease and in the probably function of the immune system in controlling cancer, or at least in postponing it to the late stages of life. It would be honest, however, to mention one influence which does little credit to my pretensions to be a wholly unbiased scientist.

Since 1958 or thereabouts, I have been prejudiced against the growing concentration of experimental work on cancer into the virus field. It is not a popular point of view with the majority of oncologists in most countries to say that the cancer viruses of the laboratory are artificial creations and that an intelligent researcher looking for a cancer virus creates the most highly selective environment in the whole history of life. With adequate ingenuity, pertinacity, and good fortune, an experimental biologist with a reasonable theory to be tested can nearly always find genuine evidence in favour of his theory, even if eventually it is disproved and discarded. This is an important aspect of the philosophy of science which it would be inappropriate to elaborate here, but it is particularly relevant when one is dealing with agents as ubiquitous and variable as viruses. If an investigator looks hard enough for a virus of cancer or epilepsy or old age, he is likely to find one with at least some of the characters he is looking for. In some ways this is an overstatement, but for me at least it is highly relevant to the problems of cancer research. I still hold that our only legitimate objective in cancer research is the understanding of human cancer, its prevention, and its cure. So I have been eager to follow every indication that human cancer can best be understood at the genetic level. With such an attitude, the association of cancer with age begins to make sense.

It is probably oversimplifying a highly complex situation to ask for a simple answer to the question: why do mice with an average lifespan of two years and men who average about seventy years both show a rapidly increasing incidence of cancer around the same ages, two years and seventy years respectively? As far as my reading goes, similar rules hold for rats, guineapigs, dogs, cattle, and horses, and I know of no mammal for which it has been shown not to hold. My answer, which may be wrong, is that in different species of mammal the rate of ageing and the rate at which cancer appears are governed by a common factor. If that should be a correct appreciation of the position, it is clear that the objective of research, whether it be armchair research like mine or the more legitimate activities of the experimental laboratory, must be to define that common factor. If we follow up the previous discussion on somatic mutation by ascribing both ageing and cancer to somatic mutation, the question changes to: how can the rate of somatic mutation be adjusted so that it is related to lifespan in both the general process of ageing and the time of appearance of cancer? Questions of this sort in various forms became central to my thinking about ageing. They gradually sorted themselves out into one. The average lifespan of a mammal is clearly as much a genetically determined quality as its size or colour. How can information in the DNA control the rate at which mutation occurs in the somatic cells of the body? Once the spark of that 'French bomb tests' committee had ignited my interest, I very soon reached that point and made it the focus of my reading.

Genes that control mutation rate

In almost every branch of biology there are one or two 'laboratory models' on which the basic phenomena can be most effectively studied. One thinks of the guineapig for tuberculosis research or the rat for most biochemical studies. In genetics one can take one's choice of the mouse, the fruit fly, or *E. coli*, according to one's requirements. But if some fine point in genetics is to be studied in detail, the choice will almost invariably be *E. coli*. It is the commonest of all the bacteria that can be grown by simple methods from the human bowel and it is susceptible to attack by many bacterial viruses. It has become conventional to regard it as the most 'typical' of all bacteria and a few chosen strains are to be found in all laboratories. Such a standard form is very easily cultivated, is non-pathogenic, and is robust in the sense that if one has an interesting mutant it is easy to keep it indefinitely with all its special genetic features maintained. More is known about the genetic structure of *E. coli* than any other organism; so much, that a few years ago Francis Crick seriously suggested as a major scientific objective, in its own way as challenging as putting a man on the moon, the experimental elucidation of the complete molecular structure of *E. coli*. The scientific consensus was that biology is infinitely more complex than celestial dynamics and that the project was far too ambitious for this century. This however does not diminish the importance of testing genetic ideas on *E. coli* and one or two of the viruses than can live at its expense. Bacterial viruses are necessarily smaller and simpler in their genetic structure than bacteria, and for some purposes they are even more suitable as a model on which to test genetic principles than *E. coli*.

It was with this model in fact that professional geneticists had already obtained the answer I was looking for, but in a quite different context from mammalian ageing. It was found, in the course of an analysis of mutants, in the bacterial virus T4D. This is no place for technical details of virus genetics. All that need be said is that it is possible to map, on the virus's circular double strands of DNA, the relative position of any gene identified by the occurrence of a mutation. Many of those genes, including No. 43, can be definitely identified as controlling a particular enzyme. No. 43 controls a DNA polymerase, an enzyme, it will be remembered, whose function is to place the nucleotide units A, G, T, and C in the right places when a DNA strand is being duplicated or repaired. A large collection of mutants of the virus, all with the mutation proved to have involved gene 43, was accumulated and put through many tests. The only one that is relevant was to determine their mutability. Bacterial viruses are rather highly mutable objects and there are simple methods of determining the spontaneous rates at which several different types of mutants appear. If it is more relevant to the experiment, a mutagen like ultraviolet light can be used to increase the yield of mutants. With such tests it was found that there were several of these mutant strains that produced many more mutants than normal. A vital point was that the mutation rate was up, irrespective of what particular type of mutant was being looked for. These mutants were classed as *mutator* strains. Even more enlightening was the fact that two *anti-mutator* strains were found,[2] which even under ultraviolet light exposure produced many *less* mutants than the normal virus. The only conclusion that could be drawn was that the normal strain—the wild type—has a DNA polymerase that, under control of its gene, makes a small proportion of genetic errors. If that gene is altered, the functional efficiency of the enzyme may be modified. For most mutants the alteration is as might be expected; if it has any effect on mutation rate at all it is towards inefficiency. More genetic errors occur and more mutants are isolated; the enzyme has become more error-prone. But some rarer changes seem to make the enzyme *less* error-prone. Taken at face value, these experiments say that the best overall situation for the survival of that bacterial virus under laboratory conditions was not that the DP enzyme should be as free as possible of error, but that it should possess a low but measurable capacity to produce a proportion of mutants.

Central to this book is the concept I have called 'intrinsic mutagenesis', which means, simply, that mutation is not directly due to mutagens but arises by error in the process of DNA duplication and repair.The degree of error-proneness, or, looked at from the other point of view, the fidelity with which the complex of enzymes that we call DNA polymerase does its job, determines how many errors will be made. It is the clock that determines the rate of mutation and the average lifespan of a mammal. Rightly or wrongly, I regard this concept as an important one, of the same general quality as the clonal selection theory. It is not really original except in its application to the problem of ageing. Many evolutionists have pointed out that there needs to be some control of the rate of mutation for evolution to proceed at an optimum rate. Particularly when circumstances demand a rapid evolutionary radiation to occupy newly available ecological niches, a higher rate of mutation may be needed than in some eternally uniform environment like the ocean depths. But in this book I am not concerned with evolution on the grand scale. I cannot escape from my medical education and it is the medical applications of somatic mutation, and all that this implies, that interest me. The concept of intrinsic mutagenesis and the error-prone DNA polymerase was clear in my mind in July 1973, and since then I have been employed essentially in seeking evidence for its validity and trying to persuade experimentalists to tool up their laboratories to follow out its implications. New evidence may appear at any time, but I do not think that I shall alter my opinion that this is one of the concepts which, once grasped, cannot be wrong.

Since mid-1973, the most important support for the theory came from two sources. The first came to my notice in a comprehensive account for physicians of the genetic disease xeroderma pigmentosum. I shall discuss this, and some other possibly related conditions, at a later stage. All that need be said here is that this has all the characteristics of a disease in which a DNA polymerase specially important in the repair of DNA damaged by ultraviolet light is grossly error-prone. The second confirmation and sign that others were thinking on similar lines was in many ways a derivative of the studies on xeroderma pigmentosum. Hart and Setlow[3] were aware of how readily differences could be shown between cultures of cells from normal human skin and similar cultures from xeroderma patients, in their capacity to repair DNA damaged by ultraviolet light. Similar cultures can be developed from any other type of mammalian skin and it was of obvious interest to use a range of different mammals for similar tests. How readily can a mouse's cells repair DNA in comparison with those of a man or an elephant? The experiments were done and gave an unexpected and dramatically simple result. If the species tested are arranged in order of life-span—shrewmouse, mouse, rat, guineapig, cow, elephant, man—the same order with no exceptions is obtained for the efficiency with which DNA damage is repaired. The mouse dying at two years repairs DNA very inefficiently in comparison with elephant and man with lifespans of approximately sixty and seventy-five years respectively.

This does not by any means prove that efficiency in DNA repair is what determines length length of life, but at least it is wholly compatible with such an hypothesis.

A final thought

This almost finishes my account of how I have become, provisionally at least, committed to one particular theory of ageing, and began to be excited about the way in which the concept of error-proneness in DNA-handling enzymes might be extended to other areas of biology. The final step may be quite unjustified, but it played an essential part in the genesis of this book. It grew out of a consideration of senile dementia and other more specifically genetic types of human dementia in which it seemed likely that these error-prone processes were also concerned. From that, it was natural to ask whether errors in the genetic control of the laying down of the central nervous system could have a bearing on the pathology of thought and behaviour. To paraphrase Arthur Koestler, there seems to be some flaw in the human nervous system which could be responsible for 'the absurd and tortured history' of mankind. One of the major socio-political controversies of the present day is in regard to the influence of genetic factors on human intelligence, temperament, and behaviour. If, as I have come to believe, the genes play a dominant part in all aspects of human behaviour, this is another area of relevance to the book's central theme. The problem of good and evil will eventually have to be restated in terms of genetics and evolutionary principles. That may be a task for the distant future, but even a simple-minded attempt to look at it now could have its uses. It is not inconsistent with a biologist's outlook to believe that only continual restatement of fundamentals in contemporary terms will serve if we are to bring meaning to modern life and fill the gap that has been left by the disappearance of the conventional religions.

4

Ageing in Mammals

The natural history of ageing

To apply the comparative approach to the human problems of ageing we must look primarily at what is known for other mammals. Experience with a fairly wide range of domesticated and laboratory animals indicates that all show a period of old age preceding natural death and that during old age they exhibit much the same tissue changes and age-associated diseases as are seen in man. Where it has been ascertained, the average lifespan for a species is relatively uniform, though the value may vary from two years for mice to around seventy years for man, and within genetically distinguishable strains of the same species there may also be quite substantial differences in average lifespan. The existence of clear differences according to species and race indicates at once that in some sense length of life is a genetic characteristic. It would in principle be susceptible to evolutionary control, always provided that an appropriate length of lifespan was in some way important for each mammalian species. That, however, is not immediately obvious and many gerontologists have expressed the opinion that ageing is something essentially accidental and quite unimportant at an evolutionary level. An apparent evolutionary significance of average lifespan might merely represent a necessary association with some other 'timed' quality of the species, for instance the age at which reproductive competence develops.

Short-lived and long-lived mammals

Most small mouse-like mammals have a lifespan of about two years, while elephants live for a time only marginally shorter than the human lifespan. Their life styles differ in many ways that are relevant to the contrast between the average length of life. Field mice and the like depend mainly on numbers for survival; they have large litters and become reproductively competent at six to eight weeks of age. There is little to indicate that their behaviour in finding food, avoiding enemies, obtaining shelter, and so on requires any form of parental training. Such small rodents are the main food source of many mammalian and avian predators and they are liable to massive destruction by flood and fire. Conversely, when for any combination of reasons normal controlling factors fail to maintain a standard population level, 'mouse plagues' are liable to arise. Gross overpopulation in some cases is terminated by the free spread of infectious disease.

In such small, rapidly multiplying rodents there are obvious evolutionary advantages to the species in a capacity to grow rapidly to reproductive maturity and to produce large numbers of offspring. It is not immediately evident why the need to die after two possible seasons of reproduction should be built into the species' genetic makeup. Two broad hypotheses need consideration. (1) The time of sexual maturation is the primary factor determined by the genes and in some way the same 'physiological clock' times the average lifespan. Or (2) the time of sexual maturation and average lifespan must maintain an appropriate relationship for some valid evolutionary purpose, although initially two distinct physiological clocks were concerned. The requirement is met by some genetically controlled interaction between the 'clocks', at present of unknown nature. There is not a great deal of difference between the two hypotheses at the evolutionary level and they need to be borne in mind only when an understanding of mechanism is being sought. The basic evolutionary reason is no doubt that in a very inconstant environment it is necessary that a large variety of new recombinants should be available to provide new genetic patterns to be tested for survival. Whenever ecological circumstances leave a large residual population, it is expedient that the older generations who have been tested reproductively should die and allow living room for the newest generations to take on the reproductive responsibility.

This, after all, is only part of the universal requirement that there should always be ample opportunity for the appearance of new gene combinations in any living population. Except under quite unusual circumstances, all individuals in nature are heterozygous, i.e. have many genes which are represented by two distinct alleles. New reshufflings to maintain genetic heterogeneity must take place at each generation. This seems to be almost the first law of evolution, and the numberless expedients that have been devised by nature to achieve this deeply impressed Charles Darwin. He wrote much about the mechanisms to ensure cross-pollination in flowering plants, and there are many equivalent mechanisms in marine animals.

The long-lived mammals, elephants with a lifespan around sixty years and man around seventy, have quite different but equally convincing evolutionary reasons for their particular lifespans. Elephants spend twelve to twenty years before becoming capable of reproduction, and during their early years require and receive care or tutelage from their mothers. They are intelligent animals, capable of dealing effectively with potentially catastrophic situations and of maintaining order and co-operative action within the group. Mature females have only one calf at a birth, and a gestation period of twenty to twenty-two months. On the average there is a spacing of four years between births. Man is the only predator to be feared by a mature elephant, and the slow rate of reproduction is under natural circumstances adequate to allow for deaths by predation on the young or accidental causes. Interestingly enough, a study of African elephants in areas with differing densities of elephant population showed that where population density was high the interval between births could average nine years as against three years with the lowest density. The universal requirement still holds that heterozygosity must be maintained and that the old must give place to the young. With intelligent animals capable of learning from others and profiting by experience, another element comes into the picture. Age brings experience in handling critical situations, especially when group action is called for. Here for the first time in the history of evolution, species survival may depend on individual qualities possessed at post-reproductive ages. In gregarious animals the herd is the most important unit on which selection plays, not the individual as such. The wisdom of an old female leader may be vital for the survival of the calves in her herd, and Medawar's dictum breaks down (see Chapter 1). Among the primates too the leader of a troop is almost regularly an old male, or, in the baboon troop, a group of two to four old males who exercise joint control.

The natural history of ageing in man

In this chapter I am concerned with the natural history of ageing in mammals, and no one who has read even a page or two of this book will be in any danger of forgetting that we too are mammals. Since the eighteenth century, Western thinkers have been writing about man in a state of nature and seeking to interpret the history and social structure of their own civilization against such a background. Over the past fifty years the concept of the hunter-gatherer phase of human culture has been developed to cover the long period before the beginnings of agriculture some 10 000 years ago. In my own part of the world a classical example of hunter-gatherer life persisted in the Australian Aborigines until less than a hundred years ago. In the highlands of New Guinea there was a typical example of tropical garden cultivation, which was probably the first step toward agriculture in hot, moist climates. Either of these groups could be regarded as living as near to a state of nature as is possible for the human mammal, and I shall take them as prototypes of natural man. Here I am concerned only with aspects of ageing that can be touched on as part of the natural history of a mammalian species.

The age incidence of mortality in either style of human living was concentrated on the young; infantile mortality was high, but a second period of danger following weaning probably killed even more children. This is the time when the child lost both the protection against poisonous, infected, or otherwise unsuitable food that breast feeding provides, and the partial protection against prevalent infections like malaria given by antibodies from the mother, first through the uterus and subsequently transferred through her milk. Accident and infection continued to cause death in childhood and adolescence, but, once adult life was reached, immunity against any ubiquitous diseases, like malaria in the tropics, had been firmly established and was maintained. Deaths from accident or homicide and from inevitable catastrophes like famine or the appearance of some previously unexperienced infectious disease would continue to occur throughout adult life.

Survival beyond fifty years was rare, and men who reached that age tended to be honoured for their experience and for their survival. They needed to be tough in every sense, and they were the natural leaders of their community. Survival gave decisive testimony to the effectiveness of their immune responses to infection, and to their strength, dexterity and courage in battle; they were the natural ones to provide the wisdom of experience in tribal matters. The old men tended to be the 'big men' of the tribe, and one of these privileged elders, even when his physical powers were failing, could retain an honoured position as a man of experience and a transmitter of traditional wisdom.

For us, the days when there was virtue in survival beyond threescore and ten are long past. Community hygiene and modern medicine allow thousands to reach an age that only one or two would attain in a standard hunter-gatherer culture. The extra quality of toughness and vigour is no longer required. Sometimes, however, one can recognize healthy vigorous old men and women in our own community who seem to have the same genetic qualities that kept the 'big men' of a New Guinea tribe at the top.

Conventional wisdom ascribes healthy old age to a variety of environmental factors, and the discovery of—or perhaps one should say publicity for—three areas of mountain country where there are undue numbers of centenarians has stimulated discussion on such matters in the last decade. The three locations are in the Caucasus mountains in the Abkhazian Soviet Socialist Republic, in the Hunza valley of the Himalayas, and a Peruvian district, Vilcabamba, in the high Andes.[1] Here there are undoubtedly unusual numbers of healthy old people, although there is scepticism about such ages as that of 168 claimed in 1973 for one of the Russians. The centenarians are found in peasant-type communities living at high altitudes, and characteristically they still take an active part in the simple everyday work of their community. Diet is adequate but very low in fats and most of the protein of vegetable origin. In the Caucasus the male centenarians appear to use a moderate amount of alcohol and are rather often described as actively appreciating the company of their fourth or fifth wife. By no means all the inhabitants of these areas are destined for extreme old age, and the most widely held view is that in each case very old individuals represent descendants of a genetically long-lived group isolated by the process of 'genetic drift'. Exceptional longevity has no direct advantage for biological survival although it may be associated with other types of vigour that have. In all probability the long-lived moiety is a chance genetic assemblage. Not infrequently, races or species of animals are found with characteristics that seem to have no advantage for survival. It is common to assume that these arise by the effective isolation of a small family group or even a single pregnant female carrying only a very small proportion of the variant genes to be found in the whole population—the gene pool—from which they derived. If such a group and its descendants are isolated from their original stock for many generations, as is particularly likely to occur in a mountain valley, the new community will express only the limited possibilities of their common ancestors. Sometimes it could include a particular genetic combination predisposing toward long life.

To put chief responsibility for the achievement of extreme old age on the genetic constitution of the individual is not to minimize the importance of the environmental factors that are usually given the credit. One must always remember that all but a tiny fraction of people die before full achievement of the lifespan to which one might say their genes entitled them. Death can come in a thousand ways, and the only satisfactory way to look at the effect of ageing is to think of it as a progressive increase in *vulnerability*. Sometimes the lethal impact of the environment may be as immediate as a bullet through the heart, and degree of vulnerability plays no part. At the other extreme we have mild respiratory infections; trivial influenza in the young is increasingly lethal in the very old and can probably be thought of as the normal end of a healthy and sheltered life.

What is needed for a healthy and secure existence can in part be drawn from the experience of our mountain centenarians. The main elements may be absence of any excess body fat associated with a simple diet relatively free from fat, and constant physical exercise, plus a sense of satisfaction in being able to carry on indefinitely with simple, socially useful tasks and in having an established status in their group with a minimum of conflict with others. Having said that, one tends to think about the not inconsiderable number of highly civilized men—Bernard Shaw and Bertrand Russell for example—who have led active public and professional or academic lives, and died, or are still healthy, in their nineties. A good set of genes obviously does not always need simple Arcadian surroundings and freedom from worry, to allow longevity.

Experimental gerontology

Strictly speaking, the nature of ageing has been studied only in man himself and those other mammalian species which it is convenient to maintain under domestication or in the laboratory. In general, domestic animals are killed at the age that is optimum for meat production, if that is their economic significance, or at whatever age it becomes uneconomic to keep them as producers of milk, wool, eggs, or offspring. Any exceptions represent human sentiment toward individual animals. Most dogs with a 'good home' are killed when clearly moribund, and racehorses of distinction, especially if that includes a long and successful time at stud, are often enough allowed to end their days in peace. For all the larger and more long-lived mammals, therefore, systematic studies of natural life-span are impractical. For most, an approximate figure can be given for the common lifespan and there is enough anecdotal information to provide examples of exceptionally long survival.

Research is a curiously human occupation, particularly when its objective is initially as vague as that of gerontologists interested in understanding the basis of ageing. For fairly obvious reasons, the first move has usually been to take a basically epidemiological approach to those phenomena associated with old age in man that can be accurately recorded. The age at death, the onset, outcome, and pathology of age-associated diseases, and measurable changes in the functional efficiency of the various bodily systems—these are the source of the conventional data from which physicians have sought by logical and mathematical analysis to answer their wider questions about ageing. If one essays an experimental study of ageing, it is clear that the first objective will be to find an experimental model, a suitable mammalian species which can be examined to see whether ageing individuals present the same pattern of senescent changes that it has been conventional to study in man.

Things being as they are in biological science, there is virtually only one option: to use the laboratory mouse. By the time an academic biologist has reached the stage of recognition and competence that will allow him to develop a specific and substantial field of research as his own, he will probably be around forty years old. This will give him twenty-five years before retirement, during which, being human, he will hope to have achieved a worthwhile advance in the understanding of his field, embodied in, say, a hundred technical papers, one or two monographs, and, no doubt extending a year or two into his retirement, one substantial volume hopefully to integrate the field in terms of modern knowledge. Individual experiments, even those involving large groups of animals, must not last too long if as is usual new experiments are to be devised to examine further implications of those just completed. Most laboratory mouse strains have a lifespan of about two years, but if one has an experiment in which 500 to 1000 mice must be followed until each mouse is dead or clearly going to die in a day or two, it will be nearly three years before it ends. For rats, the experiments would have to last four years, and for guineapigs five or six.

Ageing in mice

Serious work on ageing almost has to be done with mice. We know a good deal about how mice grow old and what diseases and pathological changes develop in their old age. Much more, in one sense at least, is known about mouse genetics than about human, and in any biological laboratory there will be available a group of pure-line stocks of types suitable to the field of study. The most commonly used strains may be mentioned mainly to indicate the sort of names by which they are known: A, C3H, C57Bl, DBA, and BALB/c. Each represents an inbred line of mice, developed by a rigid system of maintaining the central line by brother-sister mating, and, even when large stocks have to be built up, never being more than a few generations away from the central line. Within any such mouse line, one can for all practical purposes regard all the mice as being genetically identical in every respect but one: they are still either male or female.

In an uptodate laboratory too there is an additional precaution to ensure that experimental mice are uniform and that when subjected to any sort of experimental manipulation they will respond reproducibly. Science is supposed to be concerned only with reproducible experiments. We need to be sure that if two people in different laboratories do the same experiment, each using 100 mice of the same strain, both will get results that are near enough to be certified by a statistician as not significantly different. If, however, one of those experiments was quite unlike the other because many of the mice looked thin and sick, showed diarrhoea, and began to die much earlier than the other group, it would be clear that something was wrong. In addition to being of pure genetic stock, experimental mice are expected to be free of any known micro-organism that can cause infectious disease. They are 'specific pathogen free' (SPF) mice and all precautions must be taken to keep them so.

The process starts by producing 'germ-free' mice. Every normal animal, no matter how healthy, is an involuntary host to a large number of species of bacteria, and particularly in the intestines they are present in countless numbers. The embryo, or foetus in the uterus, is free from bacteria, but infection begins immediately after birth. To obtain a nucleus of germ-free mice, therefore, they must be removed from their mothers by Caesarean section and reared with full aseptic precautions so that no bacteria develop anywhere in their bodies. Even so, most of the germ-free mice will probably carry one or more types of virus, but, except for some quite unusual types of experiment, quiescent viruses in healthy mice do not introduce any serious complications—though their possible presence should never be forgotten. With proper techniques, such mice can be maintained free of any bacteria for long periods and can even be successfully mated and produce sterile offspring by the natural process in their germ-free containers. The methods, however, are far too labour-intensive for any but very small experiments, and once the germ-free nucleus is established, a group of these animals is deliberately infected with cultures of bacteria that are known to be harmless and physiologically useful to the mouse. These become the founding parents of SPF mice and they can then be handled and bred with simpler but still rather demanding techniques. To be sure that each generation of mice remains completely free of bacterial or viral disease, every precaution that can be devised to make sure that no bacteria capable of causing mouse disease can enter the stock rooms is strictly enforced. If experiments on ageing are contemplated, the precautions must cover the whole duration of the study.

There is a paradoxical flavour about carrying out experiments to tell us something about human ageing with pure-line SPF mice. Human beings could hardly be more genetically heterogeneous than they are, and from infancy onward they are constantly being infected with, on the whole, relatively mild bacteria and viruses, but few escape without at least one serious infection. Both qualities hold for every other species of mammal, wild or domesticated. Nature has gone to extraordinary pains within almost every group of plants or animals to ensure that pure-line strains can never develop. For every practical purpose in horticulture or animal husbandry, a proper degree of heterozygosity, a well-mixed set of genes, is an advantage—but not for a biologist at his experiments. Cross two strains of mice—C57Bl and C3H, say—pure line with pure line, to give what we call an F1 generation, and it will be found that the F1s will live longer and be more vigorous than either of the parent strains. Something remotely similar holds for bacteria in the gut. Disease-causing bacteria form a constant element in the web of life, but it is equally important to remember that a well-balanced population of harmless or relatively harmless organisms in the bowel is good for health and no doubt keeps the immune system in trim for the real emergencies. This holds for mice and man or guineapig. It is illuminating that a guineapig can be killed by a large dose of penicillin, not because it is poisonous in any ordinary sense of the word, but because it upsets the balance of the species of bacteria in the bowel.

Some highly interesting results have come from studies of mice, not all of them done with such flawless technique, but before discussing those results it is best to think of some of the differences between life styles of mice and men that could make it difficult to apply what might be learnt of mice to human affairs. The pure-line mouse strains of the laboratory all derive from the common house mouse that has followed Europeans around the world. White mice and other abnormally colored mutants have been kept as pets for centuries and came into laboratory use very early. The modern approach to the use of pure-line strains arose in the course of cancer research. Mice develop cancers spontaneously as they age, and it was noticed that some laboratory stocks were much more prone to develop tumours than others. Once a tumour had appeared, it was necessary for it to be transplanted to other mice if any useful study was to be made of its properties. If litter mates were used, transfer was often successful, but it regularly failed when the recipient mouse was unrelated to the one in which the cancer had arisen. If consistent results were to be obtained in such work, close inbreeding was the most likely way to be sure of uniform results. Around the 1940s, the realization of the usefulness of pure-line strains spread to virtually all laboratories in which mice were being used experimentally.

The fact that pure-line mice were developed for cancer research had an important influence on the way research on immunity developed. This arose from the use of cancer transplants to check how satisfactory the process of inbreeding was approaching the desired uniformity. For example, after twenty successive brother-sister generations of a particular mouse line, a working stock of such twentieth to twenty-second generation mice could be built up for experimental test. A tumour grown on one such mouse could be removed surgically and fragments containing only small numbers of cancer cells transplanted to twenty or fifty mice of that (hopefully) pure-line stock. If all the inoculated mice developed a characteristic tumour, it was felt that for practical purposes that line of mice could be regarded as a pure-line strain. Very soon it became clear that this ability to 'take' within the pure line had very little to do with the cancer as such. It meant simply that in animals so genetically uniform, any collections of cells, normal or malignant, could be transplanted from one to the other, and there survive and, if it were their nature to do so, multiply. From this arose everything we know about transplantation of kidneys or other organs in man.

There are excellent reasons for using the 'pure-line, specific-pathogen-free mice' for research on ageing despite the extreme difference from any human population in those respects. Everyone realizes that people die for all sorts of reasons. Some deaths are as accidental as being struck by lightning, dying of rabies after being bitten by a rabid dog, or from a perforated stomach ulcer in circumstances where no surgical help was available. Others are as completely natural deaths as that of a man of eighty-five sitting comfortably in an armchair to watch the happiness of a great-granddaughter's confirmation party and slipping out of life un-noticed with a quiet stopping of the heart. In between, an infinite diversity of ways of dying will offer a complete spectrum, ranging from the wholly accidental to the peaceful demise of the very old. A little thought will probably convince most people that natural death means a death in which the essential factors concerned were genetic ones. In the absence of accident, which covers any abnormal environmental happening, the timing of an individual's death would depend on his genes. So if we want to investigate natural death—genetic death—it is appropriate to use several strains of pure-line mice and shield them from any bacterial infection, as well as to provide the other forms of security characteristic of a well-organized laboratory.

A reasonably typical experiment in mouse gerontology could take the form of comparing the life experience of two strains, A and B, and the first generation F1 hybrid, A/B. After an appropriate breeding programme, young (four to six weeks) mice are assembled—100 A males and 100 A females—and similar groups of B and A/B. The mice will be five or ten to a cage, males separated from females; food will be sterilized and carefully arranged to contain a fully adequate diet. Deaths, either spontaneous or of mice killed when obviously moribund, are recorded each day until the last survivor is dead or when the survivor count has fallen to five or some other small number. The lifespan of each group can then be shown in survivor-time graph, which for such an experiment would be very similar to that in Figure 4.

In curves of this sort we eliminate anything equivalent to infantile and child mortality in man. The mice start the course as healthy adolescents and there are very few deaths until about two-thirds of the average lifespan. Thereafter the curve begins to bend down and at a certain age 50 per cent have died, 50 per cent still survive. This is the average lifespan, the most meaningful figure we can use to compare the longevity of two or more different groups of mice. A number of different patterns of these survival curves can be obtained and the differences can tell us some interesting things about ageing. The graphs shown in Figure 4 show, first, that there are genetic differences in longevity between strains and that not infrequently when two strains with moderate length of lifespan are used for a

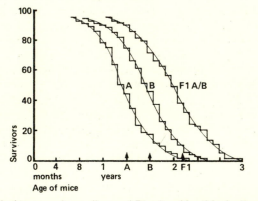

Fig. 4:   *Survival curves of pure-line and F1 mice. Monthly deaths are shown in an idealized experiment in which 100 individuals each of pure-line strains A and B and F1 (A × B) are assumed to be followed until all have died of 'natural causes'. Ordinates: number of mice surviving at the beginning of each month since birth. Arrows indicate age for 50 per cent mortality.*

first-generation cross the lifespan may be longer than that of either parent strain. This is a manifestation of hybrid vigour, a quality that has been used in agriculture particularly to increase the yield of maize. Most people have heard of 'hybrid corn'. Another point which should be made about an F1 hybrid population is that it is just as genetically uniform as the pure-line parental strains. Many workers on ageing research prefer to work with these relatively long-lived and very uniform F1 stocks. For one thing, they are just a little more like the average human population than the pure lines.

Pathology of mouse senescence

Some pure-line, and even some F1 stocks, have an unusually short lifespan. These are genetically abnormal and it has so far always been possible to identify some pathological change in one or other system. The strain AKR mice die almost regularly from leukaemia; they carry a cancer virus in the reproductive cells and in a way not clearly understood a cancerous proliferation first appears in the thymus and then breaks loose as a leukaemia which kills most in nine months or thereabouts.

Then there are some genetically unusual strains that were developed in a laboratory in Dunedin, New Zealand, and are known as NZ strains. A first-generation cross between two of these strains, spoken of as NZB/NZW or B/W mice, is highly susceptible to fatal kidney disease, particularly in the females. A population of B/W females has an average lifespan under one year, with all dead in 400 days. This is considerably shorter than either NZB or NZW pure strains, even though a considerable proportion of NZBs also show kidney disease.

From what has already been said, it will be realized that pure-line mice and F1 hybrid stocks begin to become less than uniform when we concentrate on the form taken by natural death. In virtually all curves one finds some mice dying at about half the age at death of the last survivors and, despite all our attempts to make conditions identical, mice in the same group appear to die from different diseases. A proportion will always show cancers, but, except in a very exceptional strain like AKR mentioned earlier, it will be only a rather small fraction of the population, and amongst these cancers there will be several different types. In the pure strain NZB, with which much work was done in Melbourne, there were three commonly found causes of death. The first was apparently directly due to an autoimmune disease attacking the red blood cells that was observed to some extent in every mouse of the strain. In many of them, however, the anaemia from damage to the blood cells

appeared to be too slight to be lethal. Some of these showed signs
that death was due to kidney disease; about 20 per cent had tumours
of several types, many of them probably not enough to have been a
direct cause of death; and there was always a proportion which in a
routine examination showed no obvious pathological changes.
Anyone who has studied 'natural' death in any large series of
pure-line mice, either NZB or one of the genetically more normal
strains, will have recognized that in these mice something of rather a
random character becomes evident as death becomes increasingly
closer. Death itself is scattered over the latter half of life in much the
same way as we are accustomed to see it in man. Tumours of
different types are found with increasing frequency as old age advances, but, except in specially selected abnormal strains such as
AKR, only a proportion will be affected. This holds too for the
various manifestations of autoimmune disease that may be found in
ageing mice, often including the type of kidney disease (membra-
nous glomerulonephritis) seen in the NZ mouse strains mentioned
earlier. In addition, there are signs of damage and degeneration in
the heart and blood vessels not very dissimilar in their distribution to
what is found post mortem in people who have died in their seventies and beyond.

At this stage, all that needs underlining is that there is much in the
ageing process of mice that resembles what happens in man. After a
long period of maturity where, in the absence of accidental environmental episodes, life seems to run smoothly without overt sign
of any incipient trouble, random difficulties arise and become more
and more evident until in one way or another they finish the story. It
is the nature of this irregular fraying out of the far end of the thread
of life that provides the central problem of ageing.

The interpretation of ageing

No other mammal has been studied in the laboratory with anything
like the same assiduity as the mouse. Rats too are available now in
pure-line strains, but most of the information bearing on ageing in
rats has been obtained more or less incidentally to other studies.
Probably the most widely quoted work, by McCay, deals with the
extension of lifespan that resulted from undernutrition over the
period before sexual maturity. Rats immediately after weaning were
put on a diet adequate in regard to all known nutrients but given in
amount well below what was necessary for optimum growth. At a
suitable level of diet the rats could expand their period of
adolescence to two years. Once reproductive maturity had been

reached, no further extension of life could be gained by keeping the animals on the restricted diet. Nevertheless, the overall extension of life was more than a year, and it is recorded that these aged rats showed much less chronic infection of the respiratory tract, including lungs, nasal passages, and middle ears, than is commonly found in rats ageing normally but a year younger. The results may or may not be relevant to man. In every prosperous country, improved nutrition since 1940 had led to a quite striking increase in the height and weight of children for age. It is too early yet to see whether this optimum feeding in childhood and adolescence will have to be paid for by a shortened total lifespan. There are still many undernourished children in the world, but they suffer the other disadvantages or concomitants of poverty and one can feel certain that their overall situation will shorten rather than lengthen life. Paediatricians have wondered whether the equivalent of the rat experiment should not be tried with children as a possible means of providing longer life and a healthier old age. No one, however, has grasped the nettle and attempted the experiment in an affluent community. It hardly fits with the temper of our times.

From what we know of the natural history of ageing in mammals and man, and what has been learnt from laboratory studies on mice and rats, we can feel safe in postulating the presence of a genetic control which determines the existence of the phase of old age that precedes death. This, however, gives no indication as to what the mechanism may be by which that control is expressed in the body. There have been many suggestions as to what 'really happens' to make the body grow old. In one way or another, every function of the body is involved and it is obvious enough that some aspects of ageing are secondary to others. It is natural that as the time comes for a scholar interested in medicine or biology to look toward his own old age, he should speculate about the nature of ageing. It is equally natural that in most cases he will theorize in terms of the biological discipline with which he is most familiar and that the great diversity in the symptoms and signs of ageing should allow him to make a reasonable case for his theories.

My own interest in ageing theory was initially concentrated on the likelihood that progressive failure of the immune system was primarily responsible and, as such, was no exception to the general rule. The move to a more general approach based on somatic mutation, however, escapes that criticism. To recapitulate that hypothesis more explicitly, nearly everything I have discussed is compatible with the view that the rate of accumulation of genetic error in somatic cells is itself genetically determined, and that this is mediated by the relative degree of error-proneness in DNA-repair enzymes. The short lifespan of the mouse is ascribed essentially to the fact that its genes, coding for those DNA-handling enzymes, ensure that they are more error-prone than those that are coded for by the corresponding human genes. This basic contention must be regarded as the central thread running through my whole discussion of ageing and of human diversity. It is one of those hypotheses that cannot be directly tested—though Hart and Setlow's experiment (page 42) comes close to doing so—yet is supported by many lines of indirect evidence. Only a small proportion of gerontologists, however, would offer more than a cautious interest in that approach.

Several other theories of ageing also depend primarily on a consideration of the differences between species that are correlated with average lifespan. Our comparison of mouse with elephant underlines the general finding that very small animals have very short lives. It is also well known that a small warmblooded bird or mammal needs much more food per gram of body weight to maintain its body temperature and allow necessary muscular activity. A shrew-mouse with a weight of 5 grams and a lifespan of only a year needs to eat several times its body weight of worms, insects, etc. per day. Its rate of heartbeat and respiration and the rate of its energy turnover (basic metabolic rate) adjusted for size are immensely higher than those of man or elephant. It has been calculated that the number of calories used and the number of heartbeats per gram of body weight over the whole lifespan are of the same order of magnitude for the largest and smallest land mammals, the African elephant and the pygmy shrew. We all have a foreordained quota of calories per gram, and it is up to evolution to decide how to make the best use of that quota. It is not a helpful hypothesis, but it might be hard to refute.

Another of much the same quality is concerned with the ratio of the weights of brain and spinal cord taken together with the weight of the rest of the body. By juggling with available figures, Sacher[2] produced the formula $\log x = 0.6 \log z - 0.23 \log y + 0.99$, where x is lifespan in years, z is brain weight, and y body weight, and found that it fitted most of the examples reasonably well. Large dogs with a small brain relative to body weight die around ten years, while the small-bodied, relatively large-brained Pekinese can live for more than twenty. Instintively I feel that the lifespans of elephant and man are too similar and their body weights too different to fit that formula, but I have no figures at hand for the elephant.

For those whose intellectual makeup compels them to look for an 'outside' cause of ageing, the most popular approach is to look for continuing damage by 'free radicals' presumed to be produced by oxidation processes involving fatty substances (phospholipids) in living cells. These electrically charged molecules are in principle capable of changing the structure of the fibrous protein collagen in a fashion that is characteristically seen in aged animals, and they could also damage DNA in various ways. A corroborative finding that has been given much weight by some writers on ageing is that mice treated for a long period with the substance tocopherol (vitamin E) live longer than untreated animals. This could be due to the neutralization or inhibition of the production of these oxidative changes, but the results are still far from proving that free radicals are responsible for ageing.

Ageing is an extremely complex process, involving every part of the body in a fashion that has been expressed as the progressive loss of efficiency in a dynamic system where homeostasis coexists with change. Homeostasis, which means the maintenance of the status quo by appropriate adjustments to counter any changes that may be encountered, is characteristic of all the functioning systems of the body; the classic example is how the blood leaves the skin and the muscles generate heat by shivering when body temperature falls. As an immunologist I have said that the only way to comprehend the workings of the immune system is to look at it as a homeostatic and self-monitoring system. We probably know more about the function of the immune system than about any other part of the body, and we have no illusions about its complexity. In fifty years' time our successors may come to realize that in fact the immune system is one of the easiest of the functional divisions of the body to understand. It will no longer seem as paradoxical to them, as it does to us in the 1970s, that it was only because of its relative simplicity that we could realize, as early as we did, how complex were its means of control.

Every active system or set of cells in the body is part of an automated self-regulating and self-repairing factory, equipped with an immense variety of sensors to monitor relevant conditions like temperature, blood flow, or oxygen content in every part of the body and to initiate feedback responses in a fashion as subtle and precise and far more versatile than any mechanisms yet designed for industry or for war. As we age, errors begin to be made by the sensors and the controlling circuits. All the homeostatic systems are of fail-safe type; a weakness in one aspect will call an alternative process into play, but even fail-safe systems must eventually break down as genetic errors accumulate. Symptoms and signs of functional inefficiency in such a system subject to the random accumulation of genetic error may take on almost any conceivable form, but some combinations will be more common than others. If one of the common sets of symptoms and defects can also be produced by some demonstrable effect of free radicals, or bacterial toxins, or X-ray exposure, it is inevitable that someone will explore the possibility that these more easily grasped alternatives are actual causes of the whole business of ageing. Any general theory of ageing will probably have to be couched in such general terms that its only virtue will be to help understanding of what from the biological scientist's point of view is a depressingly complex and unreproducible phenomenon. Every man grows old and dies in his own unique fashion, and anyone who has loved his dogs knows that the same is true for them.

5

Genetics of Ageing

Those who have escaped death and reach retirement age know that they are growing old. Every one of us, from childhood on, learns to recognize the stigmata of old age, and even to make a rough estimate of just how old someone is from his appearance and behaviour. Underneath the unique genetic qualities of the individual and his equally unique life's experience there is a fairly stereotyped process that we call growing old. It is so much a natural and expected part of life that it is almost unnatural to start looking for a cause of ageing.

Human curiosity—which includes scientific curiosity—is, however, insatiable, and since classical times people have written about the bemoaned the disabilities of old age and its role as the forerunner of death. Death has been the rule for all man's ancestors since many-celled organisms emerged, perhaps about 2000 million years ago. At some stage too it became necessary that organisms should die spontaneously or at least become so vulnerable to attack by predators or to other accidents that death was inevitable. Evolution and immortality are incompatible. If organisms are to improve and change, old stock must be replaced by new, and death is as necessary as reproduction. It is easy enough to see why ageing and death are universal. The scientific problem is to understand how hair grows grey, teeth fall out, and every function of the body fails progressively.

Answers will always be incomplete and must always be in terms of current knowledge in the medical sciences. There is no accepted and established teaching on the 'how?' of ageing at the present time. Possibly the most widely held attitude is one supported by Medawar, that evolutionary selection for survival does not function once reproductive performance ends; there is no mechanism to get rid of or modify deleterious genes that manifest (express) their function only after the end of reproductive life. Each species and each organ or system is therefore likely to go its own way. Nature, as it were, has lost interest and there is no reason why there should be any uniformity in the pattern of ageing in different species.

The argument has a sound biological basis and doubtless has some validity, but there is one compelling counter argument. There *are* regularities in the way all mammals age and they suffer much the same diseases of old age. Hart and Setlow's experiments showing the relation of lifespan to the efficiency with which the cells of mammalian species can repair damaged DNA have already been briefly discussed (page 42). The results provide at least a prime facie justification for seeking more straightforward biological reasons for these regularities.

The timing of old age in mice and men

Two species, mouse and man, have provided most of our information on the genetic control of bodily function in mammals. This holds for almost every field; modern immunology, for instance, has in the last decade been concentrated very largely on the genetics of the immune response in mice and the significance of certain genetic 'markers', the HLA antigens, in predisposing to immunological disease in man.

Fortunately or unfortunately, human genetics can be studied only indirectly. If people would be happy for their mates to be chosen for them by wise geneticists and to produce families of ten or twelve at request, we should no doubt learn a great deal about human heredity and produce a fair crop of geniuses in a century of research. This will not happen in the next million years. Despite this handicap, a great deal is known about human inheritance from scientific surveys of well-defined, genetically based characters, such as blood groups, abnormal haemoglobins, enzymes, and several proteins that circulate in the blood.

From the point of view of ageing, mouse and man provide almost perfect material for comparison. Mice live about two years under good conditions; some strains are longer-lived than others. Strains differ in their spontaneous incidence of cancer, but in all strains cancers are concentrated in the last months of life. It is practical for a well-funded research institute to use several thousand mice in an experiment and to be sure of completing the experiment in two years or less. It is relatively easy therefore to accumulate quite accurate information about average lifespan, the incidence of cancer in relation to age, and any other aspect of ageing in mice that may interest us.

Man has the advantage or disadvantage of living for a long time. Mortality begins to rise significantly after fifty, and by seventy-five approximately half of those who were alive at fifty are gone. The disadvantage of long life in the present context is simply from the point of view of the investigator, who, being human himself, rarely has the opportunity to follow those families in which he is interested through to old age and death. But there is a countervailing advantage. In a civilized community, accurate vital statistics are available for at least a hundred years, and in the medical literature there is a mass of information, some of which will be relevant to any clinical, genetic, or experimental investigation of ageing that may be undertaken.

From the point of view of an elementary approach to ageing, the important comparisons between mouse and man are in regard to the timing of age-associated disease, cancer in particular. In mice, as in any other mammalian species that has been studied, cancers pile up toward the end of life. In mice and men, when the incidence of cancer at each age is graphed, using logarithmic scales for both axes, both species show a steeply rising straight line of about the same slope but reaching the same high level of incidence at seventy years in man and at about twenty-four months in the mouse. The average lifespan of the mouse differs regularly from one strain to another. Most of the very short-lived strains die of some specific inheritable disease, but there is a considerable range amongst normal 'healthy' mouse strains. It becomes abundantly clear that lifespan and the rate of appearance of cancers are both genetically determined characteristics of mammals and in some way closely related.

This adds an important component to the problem of ageing. Any solution must provide a way by which instruction laid down in the DNA that carries the animal's inheritance—information in the germ-line genome in more technical terms—are expressed in action as a longer or shorter lifespan and a correlated liability to suffer cancer on an equivalent time scale. It is the inherited timing of old age that provided the stimulus to my own interpretation, but before coming to that it is necessary to look at what actually happens in the cells and tissues of the ageing individual. This leads us at once to the concept of somatic mutation.

Influence of somatic genetic errors on lifespan

In Chapter 2 I discussed briefly the nature of mutation, especially as it develops in a simple bacterium exposed to ultraviolet light. I came to the same conclusion as a majority of recent investigators in assuming that, for mutation to occur, any damage to DNA by ultraviolet radiation must be repaired and that when the repair job was a difficult one errors might creep into the reconstituted strand and distort the information it should normally carry. Under appropriate circumstances this anomalous information in the gene can produce or help to produce some detectable (phenotypic) change.

In a higher organism reproducing sexually, the situation is much more complex but similar in principle if one allows for the fact that all mammals have a double set of chromosomes and therefore two examples of each gene in their body cells. If mutation occurs in one of the reproductive cells, and if circumstances allow it to be expressed at all, its phenotypic expression will to some extent involve all cells of the body. Somatic cells will be influenced by any mutation in the germ line, but any somatic cell at any stage of development can also undergo mutation in its own DNA. Subject to their very different situation and potentialities, mutation in somatic cells is essentially the same as in the germ cells.

It is a dogma of genetics that every cell of the body contains the whole genetic information that was brought together in the fertilized egg cell, the zygote. In the plant, this can be shown quite directly by cell culture methods. With proper techniques, one can initiate a cell culture from almost any growing part of a living carrot plant. From that culture a single cell can be separated, and with infinite care and perhaps a little luck it can be coaxed to proliferate, put out rootlets, and eventually take shape as a recognizable plantlet—one can hardly call it a seedling. Once this stage is reached, the rest of the way to the mature carrot, flowers, taproot and all, is easy. Clearly, in the nucleus of the nondescript starting cell there was the whole of the information needed to construct every part of the plant.

In animals, it has proved more difficult to do this type of experiment, but something of essentially the same significance has been obtained by replacing the nucleus of a fertilized frog ovum with a nucleus taken from a body cell of a very young tadpole. Here, too, with careful handling a tadpole will develop to the specifications of the information in the nucleus of the somatic cell inserted in place of the normal nucleus that was derived from the male and female reproductive cells. As is evident from this description, all cells of the body are somatic cells except for those that have been segregated from the rest for a purely reproductive function.

Mutation is the initiating force in evolution when it occurs in the reproductive cells; somatic mutation is essentially the same process, but, since it involves somatic cells, its visible, i.e., scientifically demonstrable, results are very different. We have first to remember that cells are very small. At a rough estimate there are 10^9 cells per

cm^3 of tissue, just as there are 5×10^9 red cells per millilitre of blood. Almost by definition, mutations are rare; depending on what sort of mutation is being looked for, one might expect to find between 1 and 1000 mutants in a freshly grown culture of *E. coli* containing a thousand million cells. Something of the same order of magnitude is probably occurring in the somatic cells of the body. Where a rate of 1 in a million is regarded as high, it is obvious that it will usually be completely impossible for a single mutant cell to give any indication of its presence.

This leads to the first important practical rule about somatic mutation. A mutant cell is only going to give rise to a detectable effect if for some reason it proliferates more rapidly than the un-mutated cells of the same type in its vicinity.

In practice, this means that the only *individual* somatic mutations we can recognize and study in the human patient or the experimental mice and rats of the laboratory are malignant tumours or essentially similar conditions such as the leukaemias. Many other types of somatic mutation must occur, however, in which, although no result from any single mutational error can be observed, the accumulation of many such errors can give rise to an observable effect. These are processes which can reduce the efficiency of a somatic cell and in the limit result in its death. In all probability it is the accumulation of such changes which results in the progressive inefficiency of most organs of the body that is the cental feature of senescence.

From this point of view three areas of the genetic pathology of ageing can be separated for discussion:

(1)   The proliferative mutations that are expressed as pigment patches or tumours.

(2)   Processes may be set going which result sooner or later in the death of the cell concerned.

(3)   A variety of non-lethal and non-proliferative changes which are not recognizable individually but will diminish the efficiency of the cell to some degree.

Genetic pathology of proliferative processes

In Chapter 3 I introduced the occurrence of freckles as an example of a proliferative manifestation of genetic error in a special type of somatic cell. Here I want to go a little more deeply into that process.

One general quality of proliferative mutations is that if an end result, a cancer or a leukaemia, say, is derived from a single cell so that all the disease-producing cells belong to a single clone, this *monoclonality* can often be demonstrated by relatively simple technical methods. What we are interested in is to determine whether the cells of a tumour arise by some process affecting many originally normal cells in the region or whether all the tumour cells are descendants of a single cell in which some critical mutation occurred.

To solve the problem we must find a situation where in a certain group of heterozygous individuals the type of cell concerned comes in two distinguishable forms. Suppose, for example, that the cells in an abdominal organ were a random mixture of cells expressing one or other of two alleles differentiable by the fact that with a certain mixture of dyes they stained respectively red or blue. To the naked eye a stained section of the organ would be uniformly purple but under the microscope appear as a mosaic of red and blue cells. If tumours were common in that organ and sometimes more than one appeared in the same patient, we could prove the point that such tumours were monoclonal by staining pieces of tumour from as many examples as surgeons could provide for us. Microscopical examination of a series of such tumours would show if they were of monoclonal origin, that each tumour had either all red cells or all blue cells, never a mixture. If on the other hand all tumours had shown a mosaic pattern, it would have been certain that some diffuse process had initiated the malignant change. In fact, just such a situation is found in the common disease myoma of the uterus when it affects women of African origin who are heterozygous for a gene that determines whether a certain enzyme, G6PD, is of type A or type B. The character is carried on the X chromosome, and in any given cell only one form, A or B, is expressed. The distribution of A and B cells is random in all organs, including the wall of the uterus, from which myomas arise. Each single tumour, however, has either A or B enzyme alone and must be of monoclonal origin. Most tumours and leukaemias have now been found by such tests to be monoclonal and therefore are derived from a single mutant cell.[1]

Another important point about somatic mutation which has special relevance to malignant disease is that, just as in the reproductive cells, the active DNA of the somatic cells can be divided into two sets. The first is structural DNA, which is directly concerned with specifying the structure of a particular protein, usually an enzyme. The second is regulatory DNA responsible for all those functions of various sorts for which DNA is required but which are not mediated through a protein controlled by a structural gene. Extremely little is yet known about this 'control' DNA, but in one way or another it must be concerned with *timing*. Especially during embryonic development, changes in gene function must be made at the right time. It may be appropriate, shall we say, for the liver to start enlarging rapidly and at the same time to become a staging post where blood-forming cells may function for a while between the place of their origin in the early embryonic stages and their final home in the bone marrow. Obviously, new types of function will be called for and a whole new set of genes will need to be activated—or, as we tend to say rather clumsily, derepressed. A regulatory gene somewhere will call for action by a dozen or more structural genes and no doubt at least as large a set of other regulatory genes to co-ordinate the carrying out of the orders.

Mutation, genetic error, is, on current teaching, exactly as likely to affect regulatory as structural DNA, and there are some special reasons for thinking that nearly all the somatic mutations that we can recognize are in fact errors involving the regulatory mechanism of the nucleus. When we know more about the workings of the genome—the apotheosis of all microminicomputers—the logic of the deduction may become questionable, but, as at present, mutation giving rise to cell proliferation must involve DNA concerned in regulatory processes. Modern work on cancer points strongly to a sequence of genetic errors, somatic mutations, in a cell and its descendants as being responsible for the initiation and subsequent development of most cancers. The liklehood that most or all of those mutations involve control DNA is suggested not only by the intrinsic proliferative quality of the cancer cell line but also by three commonly seen phenomena. First, we have the common, perhaps invariable, presence of embryonic or foetal proteins in malignant cells; second, the presence of inefficient DNA-handling enzymes in cancer cells; and, third, the frequency with which completely 'wrong' proteins are produced in tumour cells, e.g. parathyroid hormone by a lung cancer. It will be more convenient to elaborate these points when cancer and other age-associated diseases are discussed in Chapter 10. Here I need to say a little on the bearing of the general finding on the possible functions of regulatory DNA in embryonic and foetal growth.

The changes taking place in embryonic development are well enough known at the simple level of recording and picturing the stages by which organs develop, and the cellular changes are visible in the microscope. Any deeper understanding seems likely to be deferred for many years, but there are plenty of what could be called unco-ordinated snippets of functional information about how proteins and other substances change in the course of embryonic life. Probably the central point in all discussion of the relation of embryonic development to cancer and other types of cell pathology is the fact that every somatic cell genome contains all the information that was in the fertilized ovum from which it arose. It is clearly possible in principle, therefore, for a mutation, a random change in a regulatory gene, to switch on a programme that was genetically appropriate to some quite different period of the individual's life history. Suppose, for example, that a cell forming part of the lining of the large bowel is by genetic error switched on to an activity appropriate to a period in early foetal life when ancestral cells were proliferating very rapidly as the bowel took on its definitive form. It would need to have been a master switch that set in motion a co-ordinated programme involving both the synthetic and the control processes needed for cellular multiplication. Any such programme activated during embryonic life would have to be under continuous control and would need to be switched off when its work had been completed. If under the very different conditions of the adult bowel both control and termination of the programme cannot occur, we can expect a condition with many or all of the qualities of a cancer of the colon. It is only to be expected that this form of unhappy accident might involve not only an excessive rate of proliferation but also the synthesis of proteins needed at the time when such proliferation was genetically proper to the individual's development.

Fallout in the brain

The first and most important example of somatic mutation leading to the death of the cell affected is to be found in the cells of the nervous system. The functional cells of the brain and spinal cord, known as neurons, do not multiply in adult life. They reach their full number in early childhood and remain active throughout life. In old age, and particularly when senile dementia is present, the number of neurons is greatly diminished in most parts of the brain. This fallout of nerve cells is in varying degrees a regular feature of senescence.

Genius, particularly mathematical genius, appears to reach its highest level in the twenties. Newton is said to have made all his main mathematical discoveries in his twenty-second year, though most of their detailed expression and publication had to wait for many years. Almost all the greatest names in mathematics—Leibnitz, Euler, Gauss, and in our own day P. A. M. Dirac and John von Neumann—have shown similar distinguished achievement soon after their teens and made their major discoveries when under the age of thirty. All remained productive for most of their lives, but the peak was in youth. To a less striking extent this holds also for people of less exalted achievement. Memory, experience, and habit can allow intellectual work to continue at an acceptably high level into old age, but in a diminishing number of people. No equivocation is possible about the effect of age on the neuromuscular functions of the central nervous system. Champions and record breakers in every form of athletic competition are in their early twenties; a thirty-year athlete is a veteran. All approaches point to a peak of physiological effectiveness of brain, mind, and ability to develop new physical skills, at about the age of twenty; for some faculties, the optimum age may be even earlier. After twenty or thereabouts, the trend of intrinsic ability is always downward, and the first demonstrable stage of senescence has begun. The simplest way of accounting for this is to postulate a progressive loss of neurons with age, and there are commonly quoted statements to the effect that all adults lose 10 000 nerve cells a day, with the implication that this is responsible for the fadeout of ability with age.

In fact there is no practical approach to follow any such progressive fallout in human beings, but it can be and has been done in rats. The results seem to be unequivocal that a steady diminution in numbers of nerve cells goes on as the rat ages. Senile dementia in man is regularly associated with atrophy of the brain and great diminution in the number of nerve cells in most areas, as judged by microscopic studies of post mortem material. This holds too for a number of genetic diseases, Alzheimer's disease and Huntington's chorea for example, in which terminal dementia is a characteristic feature. Taking an overall look at admittedly incomplete evidence, most pathologists would probably accept the view that there is a progressive fallout of neurons, which in the normal individual undergoes a slow increase in rate with each successive year, and favour the hypothesis that a pathologically accelerated, age-associated fallout is responsible for the genetically based dementias of maturity and old age.

Pathology of neuronal fallout

The pathology of nerve cell fallout points to its being intrinsic to the cell, with no indication of an infectious or inflammatory process. In senile dementia, and to a lesser extent in normal old age, there are visible microscopically two types of object, spoken of as tangles and plaques. Neither is well understood, but almost certainly they represent debris from past nerve cell disintegration. The most popular hypothesis to account for lethal processes set in motion by mutation in a somatic cell is due to Leslie Orgel[2] and is usually referred to as Orgel's error catastrophe hypothesis. He had the phenomenon of neuronal fallout in old age very much in mind in this work, but felt that it could occur in any type of cell. Our discussion of error catastrophe is modified slightly from that of Orgel and will be described in terms of what may have occurred in an elderly person dying with senile dementia. The condition is well known to be influenced by genetic factors and must in part at least be referable to changes in the germ-line genes that arose, presumably, from mutational errors in past generations. By hypothesis the genes in question must control the enzymes or other factors that ensure the accuracy with which the large molecules that carry 'information', DNA or RNA, are synthesized to their proper pattern. In the person with senile dementia, one or more of those genes is abnormal and has produced an abnormally error-prone enzyme system. In a small but always enlarging proportion of cells, errors in DNA or RNA appear and induce some secondary error in an enzyme or other gene product. Sometimes the error will involve some other enzyme important for the accuracy of protein synthesis, and this in its turn may lead to more. In a great variety of ways this process of error producing error goes on until it cascades into a lethal error catastrophe. So many enzymes are faulty or inactive that the life of the cell cannot go on. It may be clear that the greater the degree of error-proneness in the original genetically coded DNA- or RNA-handling enzyme, the greater will be the frequency with which cells develop an error catastrophe and the earlier will be the onset of the dementia by which excessive fallout becomes clinically recognizable. Automatically the process must build up with age. A healthy person will die of other causes before he loses sufficient nerve cells to become an embarrassment to his fellows, but one of the many good reasons for not wishing to see the discovery of an elixir of indefinitely continuing life is that with each decade the proportion of senile dements would rise. Whether that statement would be applicable to everyone or only to a population genetically programmed for this type of death is unknown. I suspect that the results of an investigation, if it ever became possible, would give an equivocal answer to that question.

Degenerative genetic disease of the central nervous system

An even more important problem concerns the significance of the specialized forms of dementia and of other genetic diseases of the brain which also have as their pathological basis loss of nerve cells of some particular functional type. Among these may be mentioned two examples of disease where one type of nerve cell is predominantly affected, motoneuron disease and Friedrich's ataxia; in a third type, amyotrophic lateral sclerosis and Parkinsonism-dementia complex of Guam, several regions are concomitantly involved.

In motoneuron disease the cells of the spinal cord which represent the final stage of central nervous control of muscular function are affected. They are located in the anterior horn of the spinal cord and from each a main nerve fibre (axon) passes direct to the bundle of muscle fibres it serves. Death of the nerve cell means paralysis of that portion of the muscle. In a normal individual, loss of some of these nerve cells can be recognized by detailed study of the response of appropriate muscles to electrical stimulation. Such loss begins to be evident from around the age of sixty. It is a diffuse and only slowly progressive loss of cells that must play a significant role in the waning of muscular power in old age. In patients with motoneuron disease (also called amyotrophic lateral sclerosis) the process begins in young adult life with atrophy and weakness of the muscles of the hand and slowly spreads to other groups, usually killing in two to five years from the onset. It has not been proved to be of genetic origin, but it is hard to conceive any other process which could give such a clinical and pathological picture. The presence of a component of this character in the complex genetic disease found in the Chamorro people of Guam and to be mentioned a little later points in the same direction.

Friedrich's ataxia is one of a group of degenerative changes that involve fallout of cerebellar cells and ataxia (loss of control of muscular action in walking etc.) as the predominant symptom. The conditions are inherited and symptoms appear first in childhood or adolescence. One variant, however, has its onset in adult life. Other parts of the brain and spinal cord may be involved in certain cases. The degenerative changes are progressive and eventually fatal.

Amyotrophic lateral sclerosis and Parkinsonism-dementia complex of Guam is the name given to a highly prevalent neurological disease of the Chamorro people of Guam in the Western Pacific. It is estimated that 15 per cent of adult deaths among this community are due to the disease. Suggestions have been made that it is a toxic or infectious condition, but all investigations on such lines have been fruitless and it appears to be genetic in origin. The conclusion from genetic and epidemiological investigations in 1969 was that the results were consistent with the interpretation that amyotrophic lateral sclerosis/Parkinsonism-dementia (ALS/P-D) is an inherited disease due to a dominant gene with a penetrance of 100 per cent in males and about 50 per cent in females. Penetrance is a measure of the proportion of persons carrying a gene who show signs or symptoms of its presence.

In some patients the motoneuron (ALS) symptoms appear first usually in the early forties, while in others the first signs to develop are those of Parkinson's disease, which appear at a slightly later age. Many show both sets of symptoms, and in family aggregates of cases either type may occur as well as mixed forms. Dementia develops in all and death occurs around four years from onset. The post mortem findings in the brain and spinal cord will usually show the characteristic pathological changes of both motoneuron disease and Parkinson's disease. There seems to be no doubt that a single genetic condition with variable degrees of expression is involved.

These three conditions are representative of a wide range of genetic diseases affecting the brain and spinal cord whose many and varied clinical manifestations make it almost impossible to sort them into well-defined entities. Each affected family tends to show its own individual pattern of symptoms and virtually nothing is known at biochemical or neurological levels of the nexus between the abnormality in the gene responsible and the nature of the symptoms that express the disease. Various ideas have been put forward. Probably the commonest is that the affected gene produces an abnormal toxic metabolite which acts more or less selectively on the nerve cells involved, but no experimental evidence in support has been brought forward.

In discussing the clinical, genetic, and pathological aspects of those diseases of the brain, including senile dementia, with a clear genetic background, I suggested in 1974 that we are concerned in each case with a genetic disease mediated by a gene coding for an exceptionally error-prone DNA- or RNA-handling enzyme capable of accelerating the rate at which Orgel-type error catastrophe will develop in certain categories of nerve cells. The wide range of such diseases, often localized more or less specifically to one portion of the brain, required however an important qualification. If a theory of this sort is correct, it must mean that different categories of cells have significant differences in the enzymes that make up the repertoire needed to handle the processes of replication, repair, and perhaps other DNA or RNA functions. When one considers the complexities of DNA replication and repair in *E. coli*, that implication seems to be eminently reasonable.

Ageing in fibrous tissue

Probably the most suitable example of the third process by which genetic error leads to the picture of ageing, i.e., by inducing general inefficiency and in the limit death in large numbers of cells, will be found in the behaviour of the fibrous tissues of the body.

Two of the classical signs of ageing are atrophy and wrinkling of the skin, and liability to fracture at the hip. Both are manifestations of a widespread reduction in the amount and quality of connective tissue. The whole of the body, including bone, is supported by a universal scaffolding of fibrous tissue concentrated and strengthened where necessary, as in tendons and joint ligaments. The scaffolding is essentially of fibres of two tough proteins, elastin and collagen, of which the latter is the most abundant and the one whose changes with age are best known. Collagen is synthesized and 'spun' into fibres by characteristic cells, fibrocytes or fibroblasts. Cell cultures made from almost any tissue usually show spindle-shaped fibroblasts as the predominant, often the only, cell type. Much research has been done on what happens to such cell cultures as they age, but what happens in the body's connective tissue with age must be considered first. Not a great deal is known about cell changes, but the well-marked changes in strength and elasticity of fibrous tissue, tendons, etc. with age has stimulated extensive investigation of the physical and chemical properties of collagen under various conditions.

Collagen fibres are all built up from primary collagen protein chains of four distinct types; each fibre is composed of three such chains whose types differ according to function, the commonest being formulated as α_1 I, α_1 I, α_2. The synthesis of collagen is a complex procedure; five enzymic processes are needed to make the preliminary procollagen and three more to complete fibre formation with cross-linking between the chains. Many additional processes must be involved in ensuring that the fibres are laid down in amount and direction to deal properly with local stresses, including, necessarily, movement and proliferation of collagen-producing cells. The four different types of collagen are characteristic of different tissues, the commonest being present in skin and tendon, while a distinctive type is found in cartilage.

When collagen fibres such as tail tendons from old rats are compared with those from young animals, they show an increased 'crystallinity', with more cross-linking between chains and greater contraction on heating than is seen with fibres from young animals. Essentially similar changes occur with age in collagen from man and all other mammals that have been examined. Changes may also occur in various pathological conditions. In osteoarthritis, an age-associated disease of joints, the type of collagen found in affected cartilage changes to a different type that is normally present in skin and tendon. The most likely interpretation of the chemical changes in collagen with age is to ascribe them to cumulative error involving the genes and enzymes concerned in its synthesis and laying down. Some workers, however, think of them as changes produced by chemical action of some sort on fibres that were of normal structure when they were laid down.

For fairly obvious reasons the population changes and

movements of collagen-producing cells in living tissues are difficult to study. Most recent research activity has been directed towards analysis of the changes taking place in enzymes as cells age either in the body or in cultures. Such evidence comes mainly from work on cell cultures of fibroblasts and is not directly concerned with collagen production. These researches started with the search for a suitable line of human cells on which virus for polio-vaccine production could be grown. When a promising set of fibroblast cell lines was obtained from embryonic human lung tissues, all the characters of those cultures had to be closely scrutinized, particularly to ensure that nothing suggesting malignant change occurred. In the course of such work, Hayflick, a cytologist working in California, observed that there was a limit to the number of times these cell lines could be transferred to fresh culture fluid and still retain the capacity to multiply. He could calculate that the limit was approximately fifty cell generations, i.e., after a sequence of fifty cell divisions, in the course of which a single cell could give rise to 2^{50} cells (100 million) if appropriate conditions were provided. As they approached what is often called the Hayflick limit, multiplication of cells slowed down, many dead cells were seen, and others showed visible abnormalities. Hayflick considered that this represented a form of cell ageing and suggested that the phenomenon could in fact be responsible for the ageing of mammals.

The results were confirmed in many laboratories and much work has been carried out on the changes that take place in these cultures as senescence develops. In a London laboratory[3], research had been concentrated on what happens to their enzymes as the cells approach the so called Hayflick limit and begin to show signs of structural deterioration. Holliday · and his collaborators first developed methods by which functional damage to cell enzymes could be recognized and measured. They depend on the fact that the activity or concentration of an enzyme can be measured in two ways. The first and most direct is to measure its activity against its normal substrate; if an enzyme splits starch to sugar, for example, the amount of sugar produced or starch broken down under standard conditions is a measure of the amount of enzyme by this type of assay. The second method is to estimate the amount of enzyme-protein that will combine with antibody prepared against a highly purified sample of the enzyme being studied. This is known as an immunological assay and the value obtained can be called I, as contrasted to the first 'enzymological' method with its value E. For all samples of intact unmodified enzyme, the ratio E/I should have a constant standard value. Any change in the enzyme's structure will affect its activity as an enzyme (E) before altering the chemical structures that combine with the antibody. If the enzyme in cells at the terminal phase 3 has been changed, the ratio E/I will be diminished. When such tests were applied to three readily studied enzymes, including one, G6PD, referred to in an earlier chapter (page 66), all showed a marked loss of efficiency in cells approaching the Hayflick limit.

In 1975 the same group of investigators[4] made the important discovery that in phase 3 cells the DNA polymerase can be shown *directly* to be error-prone. The method, in principle, is to set the DNA polymerase to duplicate a synthetic molecule composed only of two sorts of unit—A and T, for example. If one supplies T and A, and in addition some radioisotope-labelled G, and then tests the newly synthesized material, only A and T should be taken up. The amount of G incorporated can be measured and is a measure of the error-proneness of the enzyme. Everything found so far is consistent with what is to be expected on the classical Orgel theory of error catastrophe.

Reverting to what may happen amongst fibroblasts in living tissues, all these results would be compatible with similar changes occurring in the tissue fibroblasts producing both faulty enzymes and death of many cells. Such an interpretation is a very preliminary one and probably its only real significance is that it suggests ways by which an effective study of living connective tissue in the animal could be made and any differences between old and young animals defined. One can feel reasonably confident that within the next decade we should have a fairly clear understanding of ageing in three types of cell: the neuron in the brain, lymphocytes of the immune system, and the connective tissue cells responsible for collagen production. Between them they should provide prototypes for what can be expected in other cells and tissues.

Conclusion

In this chapter I have tried to show how an interpretation in terms of accumulating genetic error makes sense of the bodily signs of ageing in man. To the best of my knowledge of gerontology it is the only possible interpretation of ageing in man and mammals to say that it results from genetic errors in somatic cells. There may still be many different ways of looking at how those genetic errors arise.

In older works on gerontology it was possible to talk of the wear and tear theory or of poisoning produced by bacteria in the bowel, but both have become almost meaningless nowadays. The body is self-repairing, and under skilful care the most devastating injuries in a previously healthy young adult can be dealt with to give acceptable function again. Wear is only possible where repair is impossible, and in a mammal the teeth are the only tissues to which this can be applied. The teeth of an elderly Aboriginal skull of former days, with the surface worn down by abrasion nearly to the gum line and without signs of disease, represent almost the only genuine example of wear and tear associated with age.

Most of the other hypotheses represent attempts to incriminate some cause of disease, associated usually with civilized life, as the reason for ageing. As I have been careful to point out, if we consider, as we almost instinctively do, ageing in its relation to natural death, the essence of ageing is an increasing vulnerability to almost any occasion for death. If we can remedy any bad habit or chronic infection, and by so doing prolong life, we do not thereby discover the cause of ageing.

6

Immunity and Ageing

As people age they grow vulnerable, and, of all the potentially lethal mishaps they may meet, infection by micro-organisms is the most ubiquitous. Whenever an epidemic of influenza passes through a city, a sharp spike appears on the mortality curve; most of those deaths are in people over sixty, and among the old with some pre-existing illness or disability the mortality is proportionally higher. Death from influenza or any other infection has many determinants, but two stand out: the virulence of the invader, and the effectiveness of the patient's immune response.

For good practical reasons deaths from influenza are usually included under the broad heading of respiratory infections and as such have a typical curve of age incidence: high in infancy, falling smoothly to a minimum at the age of ten or twelve, and rising thereafter, slowly at first with sometimes a minor upward deviation in the twenties, and after fifty accelerating upwards with each decade. Of all the standard causes of death, the age curve for 'influenza and other respiratory infections' runs most nearly parallel to the curve of deaths from all causes. Clearly the capacity to resist infection, to neutralize toxins, and to repair damage to the tissues, all functions of the immune system, are vital for survival. Some gerontologists, notably Walford,[1] have held that the whole pattern of ageing is a reflection of the waning effectiveness of the immune system.

In one form or another I have been studying the nature of immunity since I was first told of its existence as a medical student. I have written literally thousands of pages on immunological themes and watched the workforce of people engaged on immunological research expand into the tens of thousands. And in some ways I find it more difficult to offer a short, clear, and accurate account of

77

immunity now than ever in the past. What I shall attempt is a thoroughly unorthodox approach to providing an outline of immunity for the layman.

The nature of the immune system

The function of the immune system is to maintain the chemical integrity of the body, to recognize the presence in the living tissues of anything, cancer cells or cells damaged beyond recovery, micro-organisms or alien molecules, that is not genetically proper to the body, and to initiate its inactivation and elimination.

The mechanism that has been evolved is something as complex and competent as an on-line computer, something with both information-handling and executive capacities and utterly unlike any of the human artifacts devised for such purposes. Freely moving cells, thousands of millions of lymphocytes, provide the working parts of the immune system, and many other types of cells are there to feed in information or play some other ancillary role. The recognition of foreignness, of the 'not-self' character of a protein molecule or of the chemical surface pattern of a cell, depends on surface patterns carried by the lymphocytes or, more precisely, by a rather large proportion of the body's lymphocytes.

By methods which are not sufficiently relevant to our general theme to justify elaboration, these lymphocytes are endowed with a repertoire of many thousands, possibly as many as a million, of distinct receptors, recognition mechanisms in the form of complex proteins embedded in their surface membrane. Each cell carries one recognition pattern only, although the number of lymphocytes carrying any particular pattern may number from less than ten to many millions. Recognition is by contact of receptor with foreign cell or molecule, and, once recognition has been registered, a signal is generated that calls for some appropriate action on the part of the cell. The simplest and probably the commonest response to the recognition signal is for the cell to get busy, to be activated to produce much more of its receptor material and at the same time to proliferate for a few generations. Most of the excess of receptor material is liberated into the blood as antibody, something which will attach to the specific type of foreign material that stimulated its parent cell. If that should be, for example, an invading bacterium, the liberated antibody will attach itself to bacteria of that particular type and greatly facilitate their removal by scavenger cells. Other types of cell, called T lymphocytes because they are generated in early life from the thymus—T for thymus—have other sorts of receptor, to detect one particular group of not-self patterns. They are specially fitted to detect foreign chemical configurations on the surface of another cell and recognition is followed, either directly or after the recognizing lymphocyte has proliferated, by a cell-to-cell attack that ends in the destruction of the foreigner.

Each of those immune responses, antibody production by B lymphocytes—B for the bone marrow from which they arise—or direct cellular attack by T cells, is much more complex than I have suggested in that outline. In every one of the last five years, at least a thousand good research papers on the behaviour of B cells, and more on T cells, have been published. Most of them have dealt with ways by which the numbers and the activities of the B and T lymphocytes are controlled under various circumstances. Like any modern factory, automated internal control to maximize efficiency to the overall benefit of the economy is the rule.

Control of the immune system

The immune system is a homeostatic and self-monitoring unit.[2] That is probably the most compact way of expressing the nature of its internal control. Homeostatic is a useful word, which should be more frequently borrowed from physiology for general use. It refers to the way by which the status quo is maintained by appropriate adjustments whenever unusual activity or emergency threatens to push things off balance. If, for instance, blood is to be adequate for muscular work, it must be supplied to muscle at the right pressure, with the right amount of oxygen being carried by the red cells and at the standard temperature. Many other specifications at the chemical level must also be met, but those three will serve as examples. Anyone with a little knowledge of physiology will know how essential it is that blood pressure stays at the normal level, that temperature hovers around 37°C., and that oxygen is always passing to blood and tissues. All are homeostatically controlled.

Similar requirements hold for the immune system, but here the need is to be able to handle emergency situations rather than to maintain a status quo, and a correspondingly more complex and flexible system of control mechanisms is required. The central requirement is that ways must be available for the system to 'know' how effectively each functionally distinct type of lymphocyte is dealing with the problem on hand and to intervene effectively as required.

For this to be possible, a highly complex network of interactions between cells is called for. In all sorts of combinations a receptor on one cell makes effective contact with a chemical pattern (often called an effector) on another; and a signal tells one or both cells to react appropriately to the situation, to proliferate, to die, or to make some specialized response. Details are unimportant. What is vital is that the great majority of those interacting receptors and effectors are laid down genetically, and if the system is to work satisfactorily they must be free of error. Again, like the more sophisticated kinds of automatic control in petrochemical industry, there are controls within the immune system that can deal with partial breakdowns of other controls—what can be called a fail-safe system. An 'attack of measles' in a non-immune adult, with fever, rash, pains, headache, and general misery, represents a major crisis to the immune system, with great destruction of lymphocytes and the temporary disappearance of some types of immune response that had been present before the attack. But in a day or two the symptoms disappear and within a few weeks tests of the immune system show that everything has been brought back to normal and a permanent immunity to measles stored away in the computer's memory.

The internal genetic control of lymphocytes and the stem cells which represent the link between them and the fertilized ovum from which everything in the body takes its origin is extremely complex. So many functions need to be distributed amongst the working multitudes of lymphocytes that a higher proportion of genetic errors must be expected than in cell lineages with less demanding responsibilities. Disease based on genetic error in lymphocytes of the immune system is common, particularly in later life. The common childhood leukaemia and the typical old-age leukaemia—chronic lymphatic leukaemia—are well known. There are tumours, such as lymphosarcoma, arising in lymph glands, or multiple myeloma in the bone marrow, and above all there are the autoimmune diseases that in the last few years have begun to be understood as a result of the appearance of clones of cells which combine a capacity to react with some normal component of the body and a resistance to the normal process by which such cells are recognized and eliminated. As a result the cells may proliferate in almost cancerous style, and, depending on the autoantigen (the normal component with which they interact), various forms of serious disease may result.

The self-monitoring function

When the system threatens to become too elaborate, special arrangements are necessary to deal with errors and breakdowns. In descriptions and advertisements of the new miniaturized computers, the ease with which a faulty part can be detected and speedily replaced with another 'little black box' is emphasized. One doesn't mend, one replaces components. Similarly in the immune system, many immunologists believe that potentially malignant or autoimmune lymphocytes are appearing frequently in all of us but are so effectively dealt with by what I have called the self-monitoring function of the immune system that we are never aware of their existence. One of the most interesting proofs of this comes from the frequent occurrence of lymphocytic cancers in patients whose immune response is being suppressed by drugs and in children who are born with a congenitally deficient immune system.

A patient who has received a kidney transplant will only retain the grafted organ if the natural immune reaction to reject it is damped down by appropriate *immunosuppressive* drugs, such as 'Imuran' and cortisone derivatives. Such drugs will naturally also affect other useful immune responses, and it is one of the unhappy features of what in more than 70 per cent of patients is a successful life-prolonging operation that about 1 per cent of the patients develop malignant lymphocytic tumours, often in the brain. These tumours are about 350 times as frequent in these patients as they would be in normal people of similar age and sex.

Such tumours are even more frequent in children with one of the genetic diseases that cause deficiency of the immune system of a chronic character. Children with either of two such diseases—rarities that few have heard of, called Wiskott-Aldrich syndrome and ataxia telangiectasia—die not only of lymphocytic tumours but also of some commoner types, such as stomach cancer. Having regard to the fact that both types of cancer are very rare in children, there can be no doubt that the interference with the immune system is directly responsible for the appearance of the cancers.

The two sets of results, with immunity damped down with drugs or inadequately developed for genetic reasons, are concordant in suggesting that rather large numbers of lymphocyte cancers are initiated at any time of life, but that as long as the immune system is in good shape the dangerous cells are recognized and eliminated by the self-monitoring process I have described. The more ordinary tumours of skin, stomach, and uterus show a definite but much less striking effect of the same character. I was responsible for introducing the idea of immune surveillance a good many years ago, but it is an idea that has not been uniformly accepted. Most immunologists are sceptical whether there is any significant capacity of the immune system to interfere with the development of epithelial cancers like those of skin, stomach, intestine, lung, and uterus, although they fully accept the reality of effective surveillance over lymphocytic cancers. On the other hand, laboratory workers on cancer as well as surgeons specializing in that field seem to be impressed with the probability that an immune response is often effective in getting rid of small numbers of cancer cells unavoidably left behind after surgery. Mathé and others have claimed that if appropriate stimulation of an immune response against leukaemic cells is arranged to follow standard treatment by a mixture of cytotoxic drugs, prolonged remissions are more frequently obtained in children with acute leukaemia.

However, this interlude of discussion about cancer and immunity is only directly relevant to the general quality of the self-monitoring function of the immune system, and it is expedient to move to an aspect that is even more significant for ageing, namely, autoimmune disease.

Autoimmune disease

If a biochemist with a reasonably broad knowledge of his subject was suddenly confronted for the first time with the capacities of the immune system, he would probably be quite incredulous. It is fantastic how the immune system, while accepting all the hundreds of thousands of the body's own components as proper and to be tolerated, rejects or develops a capacity to destroy any cells or proteins that differ even trivially from 'self'. It appears too that despite the general rule of rejecting foreign cells there is a most important exception, the embryo and foetus in the pregnant uterus, which *must* be tolerated. The subtlety of these various discriminations underlines what I have said about the intricacies of control within the immune system and prepares us to accept as inevitable that mistakes in differentiating between 'self' and 'non-self' must sometimes occur. And it is wholly in line with our expectations that such mistakes do occur, and when it is impossible to rectify them gross disease, autoimmune disease, may follow.

During the last ten years of my career as an active research worker, I revived an interest in clinical medicine by a close association with the physicians, Drs Wood and Mackay, of the Clinical Research Unit that was the joint responsibility of the Royal Melbourne Hospital and the Walter and Eliza Hall Institute. It has been interesting to watch how more and more diseases of previously unknown origin are now known to be associated with the damaging action of cells of the immune system on other cells of the body. As is the rule in modern medical science, autoimmune disease is far more complex in its causation than we thought ten or fifteen years ago. A very large section of what we used to call diseases of unknown cause are now placed in the list of autoimmune conditions. It is only right, however, to mention that there are still many physicians who persist in looking for some more easily comprehensible cause, preferably a virus, in such cases and regard the anomalous immune reactions as secondary to the real, still undiscovered cause. My own conviction that autoimmune disease is essentially genetic and somatic genetic in origin has been greatly strengthened in the past two years with the recognition of the significance of tissue-typing in relation to disease.

Everyone with a smattering of medical knowledge knows that it is a simple matter to graft skin from one part of an injured person to make good a burnt or denuded area elsewhere, but this cannot be done with a graft from another person. Investigation of this phenomenon brought to light immune differences between individuals which could be detected by suitable blood tests and which determined in part whether or not a graft would take. It has been the general experience that if a patient needing a kidney transplant has a brother with exactly the same tissue type—and the chance of finding such a sibling is a little less than one in four—one can almost guarantee that the patient will accept his brother's donated kidney with a minimal amount of trouble.

Each person has a pattern of four HLA antigens, divided into two sets, A and B, of which an A + B set is present on both examples of the responsible chromosome. There are now more than forty such antigens known—as well as two other groups, C and D, which need not be considered here. For historical reasons the numbering of the antigens is irregular, as between A and B; some important antigens may be mentioned: A1, A2, A11, and B7, B12, B27. A typical formula for an individual would be A1 B8 on one chromosome, A9 B12 on the other. To work out those HLA types is an interesting and rather tricky laboratory manipulation and the results are of obvious importance for medicine. Now that the techniques have been standardized, large numbers of these tests are being made in almost every medical school and research centre in the world and a great mass of information has been accumulated.

Out of that information has come the general statement that most diseases affect people of any tissue type indiscriminately, but that types B7, B8, and B27, and only those types, predominate in certain chronic diseases, all of which have been considered for other reasons as being predominantly autoimmune in nature. I give the list, fully recognizing that many of the names will be unknown to most people. Type B7 is in excess for multiple sclerosis. Type B8 for Graves' disease of the thyroid, Addison's disease of the suprarenal gland, early onset diabetes, myasthenia gravis, chronic active hepatitis, dermatitis herpetiformis, and coeliac disease, which is due to an immune reaction of abnormal character against a protein of flour.

Type B27 for ankylosing spondylitis (poker back), acute anterior uveitis (an eye disease), juvenile rheumatoid arthritis, and a group of conditions in which an infection triggers off a 'reactive arthritis', an inflammatory condition which does not represent an actual infection of the joints.

From our angle, what is important is that this regularity means that genetic factors are important in all these diseases and that the genetic influence works through the immune system. With this background, one can state the requirements for an autoimmune disease to appear as:

(1)   A genetic predisposition centred on the immune system.

(2)   A somatic genetic error that renders a lymphocyte resistant against the processes that should be able to destroy it as part of the immune system's normal control. If the lymphocyte type in question recognizes and reacts with a normal body substance, the stage is set for autoimmune disease. As in most forms of cancer, the monoclonal character of several forms of autoimmune disease can be demonstrated and so provide direct evidence in favour of this requirement.

(3)   Accessible amounts of the autoantigen, i.e., the substance that can stimulate the changed lymphocyte to multiply.

(4)   A 'trigger', often a minor infection, which in one way or another sets the process in action.

Immune defects in old age

With this outline of the immune system and how it can go wrong and sometimes facilitate instead of preventing disease, we can come to grips with what this chapter is really about: how the progressive accumulations of genetic errors in the immune system are responsible for many of the manifestations of old age, including the characteristic vulnerability of the old.

I have said something about influenza as being notorious for the way in which an epidemic, particularly in the days before antibiotics, took heavy toll amongst the old. Only some infectious diseases show this characteristic. One never hears of an epidemic of measles killing old people, for instance. There is a famous episode in the history of epidemiology, when in the early nineteenth century measles was brought to the Faeroe Islands in the north Atlantic, an isolated outpost of Danish civilization whose last epidemic of measles had taken place sixty years before. Measles spread rapidly through the whole community, affecting everyone except those over sixty who had experienced the last epidemic as children. They had kept their immunity intact.

It is the rule that high mortality in the old from infection is seen when the micro-organism responsible is one the community has not previously experienced or when, as with influenza, the virus changes its immunity-producing quality, its 'antigenic type' every few years. In a real sense, any extensive and severe epidemic of influenza is a 'new disease'. The commonsense conclusion is that old people die from infection because their immune systems have lost their former capacity to respond briskly and generate a long-lasting immunity against most of the infections they encounter.

Laboratory tests in mice and other animals show that capacity to produce antibody or to mount some other type of immune response is at its height when the young animal is just approaching maturity and falls progressively after the halfway point of the average life-span. In the human species, *optimum* immune performance is, as I have said previously, at around ten to twelve years of age; for a few years around eighteen to twenty-five, particularly in young men, the immune responses may be too strong and occasionally kill by their violence. After that, immunity in most people remains effective at least until the fifties. Serious deficiency in tests of immune responses is rarely seen in healthy old people until after seventy-five, but the tests that are used almost certainly will miss minor degrees of immune inefficiency.

Most of the dramatic examples of autoimmune disease occur in relatively young people, probably because they require the presence of major genetic anomalies. The autoimmune conditions associated with old age are common, but usually of low intensity. Many immunologists suspect that much of the joint stiffness and rheumatic troubles of the old are autoimmune in origin, but this has not been established. Autoantibodies of the sorts that are commonly tested for in the laboratories become more and more frequent with age, but they are not usually associated with recognizable disease in the corresponding organ. They must probably be taken simply as a sign that errors are piling up at random, but for the most part are producing patchy inefficiencies rather than visible disease. Nevertheless, evidence is accumulating that if people over eighty are tested for autoantibodies, those who show them are significantly more likely to die within the next two years than those without them. The errors may not be as trivial as we thought. One suggestion is that autoantibodies have some indirect correlation with damage to the circulatory system by autoimmune processes.

Two relatively common diseases in old age are chronic lymphatic leukaemia and multiple myeloma. They represent ways in which lymphocytes as such, or their plasma cell derivatives, have escaped from control. In each case the cells which throng blood or bone marrow in their thousands of millions are derivatives of a single cell in which some exceptional error has set it proliferating almost without restraint to produce an enormous descendant clone. The process is almost equivalent to a cancer of mobile cells and most cases are fatal after some years of illness. Both diseases are strictly age-associated in the sense that in each successively older age group the percentage of people with either condition increases. If one includes a condition closely related to myelomatosis but in which tumours are not produced in the bone, 3 per cent of men over seventy-five have disease that produces an excess of one (monoclonal) type of immunoglobulin or antibody, and the percentage is still rising at ninety.

Much has probably still to be learnt of the interrelationships of ageing and immunity. There are insistent hints that the greatest killers of all, the cardiovascular diseases, especially atheroma and hypertension with their lethal consequences from heart attacks and stroke, may have an autoimmune component. They are, however, still only hints, and a fair case can be made for any immune anomalies being secondary to degenerative changes in the arterial lining or elsewhere.

Another feature of the pathology of old age, something that is found at post mortem without producing obvious symptoms during life, is the deposition in various tissues of an abnormal protein, spoken of as amyloid. In the old days of chronic sepsis and tuberculosis, when a patient died after years of illness, great masses of amyloid might be found in liver, spleen, and kidney. Nowadays we don't see such cases, but if one looks carefully at properly stained tissues from almost anyone who has died after reaching seventy, small patches of amyloid will be found here and there, even in the brain. Especially following a period of senile dementia before death, amyloid plaques and 'tangles' of broken down nerve cells are likely to be seen in many parts of the brain.

Recent work on amyloid has shown that in many cases it is made up of tiny fibres, some of which are undoubtedly derived from antibody protein. Other components also appear to come from cells or blood proteins that are concerned with immunity; quite often, damaged lymphocytes and plasma cells can be recognized at the edges of the amyloid patch or plaque. It makes sense, therefore, to suggest that the patches represent little graveyards, marking sites of conflict, as it were, between autoimmune cells or autoantibody and some 'target cell' toward which they were directed. The presence of amyloid plaques in the brain is specially interesting, as with the somewhat doubtful exception of multiple sclerosis no autoimmune diseases of the brain are known. If senile plaques in the brain substance do represent relics of autoimmune activity, this may become a lead to the understanding of senile dementia or psychosis. For the present, other interpretations are more popular, but we could see interesting developments in the near future.

7

Death

In the end we die, and for us it is as if we had never been. To the individual, death is the same nothingness as existed before mind began to dawn in infancy. Death may be no more than the direction in which time's arrow points and, once it has been achieved, is meaningless. We face the extinction of thought every night without a qualm. Logically we should have the same attitude to death, but evolution had other views. It is expedient for the survival of every species that at least as long as the possibility of having offspring exists, every individual organism should struggle to survive. Even when that possibility is long past, the pattern has been too deeply etched to be wiped away. I can look forward with a calm fatalism to the oblivion of death, but I shall undoubtedly resent and struggle against the inevitable when the process of dying begins.

To me it is just unbelievable that any individual with a background of physiological and clinical knowledge could begin to think of consciousness and behaviour in the absence of a central nervous system functioning with all its ancillary services. Death stops all brain function, and if there should be mind-stuff out of which consciousness is constructed by our neural circuitry, that too will lose its structure, just as in a few days or months the molecular structures of the body will be dispersed formlessly through the biosphere.

In this chapter I want to write about death as if, on the one hand, I knew it to be something quite unimportant, something to be grateful for as an escape from the slings and arrows of a harsh environment; but, on the other, to understand the fear of dying, to sympathize and look for ways of making the last days more tolerable. Death is the culmination of maturity and old age, and I suspect that our reactions to the actualities of dying are determined by the genes we were born with to the same degree as any other aspect of our behaviour.

The physiology of dying

When a sudden loss of blood pressure occurs, as in fainting, shock, or a large haemorrhage, the blood supply to the brain is reduced to a level that is inadequate for some or all cerebral function and corresponding symptoms appear; loss of consciousness, usually temporary and rapidly restored with re-establishment of the circulation, is the most conspicuous of these. The more deep-seated visceral functions, like respiration and the heart beat, continue, but if blood pressure continues to fall and there is grosss deficiency in the supply of oxygen and other metabolites to the brain, all functions, including the visceral ones, fail and the individual dies. Irreversible loss of consciousness is a major sign of death, but of course it may precede death by many months if the patient is held in the intensive-care ward of a modern hospital. In the absence of modern technical intervention, death, in the sense of permanent loss of consciousness associated with loss of the basic functions of respiration and blood circulation, is always due to failure of brain function in those centres that are essential for the visceral functions. Sometimes the visceral process will fail first and the shutdown of brain function follow immediately, or it may be in the opposite direction, when a haemorrhage in the brain stem forces a sudden closedown of everything.

That very orthodox outline of the process of death has developed some queer implications in this modern age. It is probably almost true that a modern resuscitation and traumatic surgery team could take a head as it rolled from the guillotine and bring it back to some sort of consciousness. The techniques of intensive care can in principle be applied to all the standard acute catastrophes that can bring injury or disease to a lethal termination. This can include, for example, acute coronary blockage, pulmonary embolism, ruptured aneurysm, obstruction of the larynx, and so on, and if what is needed is done soon enough and skilfully enough the patient will survive, at least temporarily. The circulation of blood can for a time be dealt with as well by a 'heart-lung machine' as by the heart; mechanical respiration, when the patient cannot breathe, is commonplace and if needed can go on for months. Where money and skill are both unlimited, death can nearly always be postponed for a while.

The more dramatic life-saving possibilities that I have mentioned apply for the most part to bodily catastrophes occuring in a relatively well-functioning organism. It is fortunately almost unheard of to go to such extremes with a person in the throes of natural death. Nevertheless, what the press told of medical and surgical activity with thirty consultants around the deathbed of General Franco in November 1975 may be a foretaste of what the great can expect in future. It also adds confirmation to Ivan Illich's thesis in *Medical Nemesis*,[1] that what he calls the medicalization of society has brought the epoch of natural death to an end. However, for the ordinary citizen whose time comes in his seventies, prolongation of life will probably be sought by less heroic but probably still uncomfortable means. Death is always depressing, always uncomfortable when it is not subject to chronic unremitting pain, undignified, and lasting too long. Bernard Isaacs of Glasgow has written of what he calls pre-death; it is a retrospective name for the period which elapsed from the day the patient became finally incapable of looking after his or her bodily functions and was thereafter inescapably dependent on the care of others. It is depressing to find that on the whole the longer the individual lives the longer will be the period of pre-death. In Glasgow women dying at eighty-five or over, it averaged a full year.

Physiologically this slow dying in the elderly must represent continuing degeneration and increasing vulnerability with disabilities reacting to increase other disabilities. Bed sores with low-grade infection result from immobility, and infection increases degenerative changes elsewhere. One reads many accounts of the effects of mental attitude in determining the time at which death actually occurs, with its implication that the dying can sometimes consciously resist the end for a limited time, or, rather precipitately, give away all desire to live, and die, as it were, deliberately. As an Australian, I have been aware since childhood of the story that among Australian Aborigines living in traditional tribal fashion, medicine men could kill an individual by 'pointing the bone' at him. In recent years there have been a few accounts by European physicians of instances they had observed that appeared to conform to that pattern. An Aboriginal, convinced that a death spell of this sort had been laid upon him, would become quiet, apathetic, and eating little; and die for no medically evident reason within a few weeks. An Australian surgeon, G. W. Milton,[2] writing in 1973, felt that he had seen a similar phenomenon in a proportion of patients of European origin with malignant disease. It followed their first full recognition that they were suffering from incurable cancer. The immediate response was likely to be a brief period of blustering euphoria, then a lapse into apathy, and early death for no definable reason. One may believe that such things happen without worrying too hard to imagine by what mechanisms they may be achieved.

Much has been written about the changing impact of death on social life and traditional attitudes toward it through the centuries.

There has never been a simple acceptance of the biological inevitability and naturalness of death, and this holds as much now as
any time in the past. I have no interest or competence to follow the
various fantastic attitudes to death through classical times and
mediaeval Europe, but I am interested in Illich's diagnosis of the
present situation with its concentration on the ritualization of the
process of dying, with patient and doctor interacting in their expected but often illogical roles. Without being at all sure that Illich
would concur, one may first look at what seem to be the objective
needs for human health from a logical cost-benefit point of view.
They are:

(1) a standard of living that will allow cleanliness, pure water,
and adequate food;

(2) skilled technical help in preventing infectious diseases and
malnutrition, particularly in the very young;

(3) medical care that will facilitate recovery from disease and
disability due to the environmental impacts of trauma and infection;
and

(4) the provision of orthopaedic surgery and prostheses for those
who are genetically handicapped or suffering the physical disabilities of misfortune or old age.

Implicit in that summary is the understanding that the end result
is a healthy human being, capable of taking his share in the work of
the community. Biologically considered, nature has no interest in an
organism that is significantly deformed or inadequate to meet the
needs and dangers of the environment. For the human invalid, help
must come basically from human compassion, especially of kinsfolk.
I have some sympathy for the point of view that the incidental aches
and disabilities of old age are better dealt with by the unsophisticated remedies and affectionate support of friends and kinsfolk
rather than by physicians and professional paramedicals and welfare staff. This is an approach that goes right back to the hunter-
gatherer phase and it has its attractions; but it is an attitude that
cannot persist in a literate, affluent Western society—where the right
of everyone to medical treatment at the state's expense is axiomatic.

The outstanding result of state health policies everywhere has
been a steadily increasing demand for medical care and a much
more than proportionate increase in its cost. There are potential and
actual advantages for health in a rising standard of living, but at least
as much potentiality for positive damage to health and survival. The
influence of alcohol, tobacco, and lack of exercise in shortening life,
particularly in males, is statistically established, and along with other
aspects of civilized life they are doubtless responsible for many of
the ailments, minor and major, for which medical help is sought. An

overworked general practitioner can survive only if he makes time for his patients with real problems, by presenting some appropriate tablet or capsule to provide symptomatic relief for the complaints of the rest. It is the only practicable thing to do, but it is not good for the health of the community or for the self-respect of the doctor. The need for 'a bottle of medicine' to terminate the consultation was accepted for centuries, and until the last half century probably did little harm. Present-day drugs are both more effective as symptom blockers, and more dangerous. I have heard at least two geriatricians claim that the best first approach for handling an elderly invalid is to stop *all* his pills and see whether he does not improve immediately!

Iatrogenic disease is a reality, but one must always be prepared, in any particular case, to balance the initial need for action and the expected benefit of the treatment given by the doctor against the ultimate unfortunate effect of his handling of the patient. All this, however, is merely leading up to the medical aspects of the process of dying, particularly the role of the professional attendants.

From the earliest stages of human development one can confidently assume that fear of death was well understood and frequently in the forefront of consciousness, and that rituals had developed to channel the emotions provoked by the imminence of death in a kinsman. It must have been an evolutionary necessity for large-brained mammals that their response to any threat to life, whether from predator or human antagonist, should be reinforced by the visceral processes that in man reach consciousness as fear, anger, and the like. The universality of those human reactions to danger must have played a major role in shaping the various myths and social rituals about death, including our own, but it has also the more substantial and logical result of calling for all the real or presumed safeguards against death that can be devised. This necessarily includes the provision of medical care. Where critics of the medical profession are probably right is in their assessment of the potential dangers of using conventionally accepted types of therapy for conditions that are usually far from being fully understood. It is inescapable that at every stage in the historical development of clinical medicine it has been, is, and will be necessary to modify the attitudes of both doctor and potential patient toward a better assessment of what are the best contemporary principles of medical care.

The doctor's role and limitations

One of the first things that one learns for oneself during clinical training is to recognize when a patient is 'really sick' and must be admitted to hospital. The maxim 'when in doubt never take positive action' is anathema to many physicians and will undoubtedly result in an occasional tragedy, but on the balance I am certain that it would greatly diminish the amount of physical suffering and unhappiness in the community. It should hold for both sides of the clinical encounter. The need for medical help in serious acute disease or injury and for any appearance that may represent a malignant growth is obvious. The problems arise in the chronic ailments, particularly of old age. For these, the decision whether or not to call a physician will be determined by the patient's temperament and his response to social conventions and pressures. It is for people with such troubles that orthodox medicine tends to be least effective and most prone to induce iatrogenic disease, and where the various forms of folk medicine, religious healing, and miscellaneous quackery are liable to have at least temporary success.

The doctor, on his side, has a much more difficult set of decisions, and in the present state of public opinion and of his liability to be sued for malpractice it is probably impossible for him to avoid prescribing unnecessary drugs or recommending surgery that is almost certain to be ineffective. It worries me, as it did Lord Platt, that almost all modern clinical research is within the areas of chronic degenerative disease, where not much can be done and where excessive investigation and treatment involving discomfort, indignity and pain does little more than fill the interval to death. The sum total of satisfying life would probably be increased if, when a condition is recognized that has less than a 50-per cent chance of definitive cure, treatment should be limited to what is necessary to offer comfort, irrespective of its effect on length of survival. How often does one, at a friend's or a relative's funeral, feel sad that death had not come months earlier. If we can shorten that pre-death period—the time that elapses between the day when the patient first became continuously dependent on fulltime care and the day of his death—it will be better for all concerned.

At the present time the predominant attitude is that every individual is entitled to go on living as long as is possible at whatever the cost to the community or to himself and his family, and that the primary objective of medical care within the welfare state is to let this be achieved. That climate of lay and medical opinion is almost wholly responsible for the insanely mounting cost of medical care. Despite this expenditure, everything that I have read would indicate that over the last decade it has produced no improvement in the health of the community insofar as this can be assessed by the demand for medical help or the consumption of drugs. There is not even much evidence that it has significantly increased the average expectation of life of people in their sixties and seventies. It is well to remember that if all deaths from cancer were eliminated, the average lifespan would increase by only 1.5 years, and that, in some studies at least, the provision of intensive-care wards has had no statistical effect on survival of acute cardiac patients. I have said on other occasions that the full requirements for rational medical care had been achieved by 1955 and that the main task of medicine now is to see that good 1950-5 medicine is available throughout the world. It will probably never become possible unless the attitude to death both of laymen and doctors in the affluent countries is greatly changed.

The eventual inevitability of his own death is recognized by every mature individual. He knows too that, so far as earthly matters are concerned, *nothing* is of any significance to him once it has occurred. Only the memory of his abilities and affections in the days of his mature competence means anything to those who knew him. The memory of the phase of pre-death is wiped out, even by those who resented the labour it entailed for them, and the deceased individual is forgotten, like the billions who died before him. There is a nearly universal taboo against the discussion of death; even the word is avoided in favour of some acceptable alternative whenever possible. As many have said in recent years, the time seems almost ready for that taboo to be lifted in the same way as the taboo against the public discussion of sexual matters has been over the past two decades.

It is at least arguable that much of the current taboo on any discussion of death in its personal implications is associated with the ritual of what Illich calls medicalized death. Surely it will eventually be possible to avoid, as he says, the present situation, where socially approved death happens when a person becomes useless, not only as a producer but also as a consumer. As a first approach there is one simple measure anyone can take who is over sixty-five. He should carry with him a signed request that if he is found unconscious from any cause, only traditional unsophisticated methods should be used in any attempt to revive him. I do this myself, largely because of the liability that irreparable brain damage in the old may result from any significant period of oxygen lack, but also to avoid the necessity of, in effect, dying twice. The desirable corollary, of course, is that the same attitude should be taken by those who make the decision as to treatment of the unconscious old. That may still be several years away, but it is so reasonable and so much in accord with what many people feel, that it must be the most vulnerable point at which to start breaching the tradition that life must always be prolonged as far as modern technology will allow.

It is probably justifiable to reiterate here the suggestion made earlier, that an inheritance from the hunter-gatherer phase of human evolution is largely responsible for the virtually universal public stance in Western democracies that, once a child is born, it is a universal imperative that, at whatever cost, that child's life must be conserved until the resources of medical technology are exhausted, whether the period is a few hours or through life to extreme old age. Biologically, this is a thoroughly irrational attitude which, however, is enshrined as one of the basic human values and is something to which every politician, preacher, or publicist pays lipservice. Illogicalities and paradoxes abound in this area of human behaviour and human protestations. Many have commented on how another inheritance from the hunter-gatherer phase finds every sovereign state committed to training a substantial and often very large proportion of its young males to kill other young males similarly trained elsewhere. In this field the taboo against killing vanishes completely.

The value of human life

In the overcrowded world that is already upon us it will be logically absurd to continue to believe that all human life must be conserved at any cost and to maintain this as one of the absolute unchangeable principles of human society. The real need is to look at how best a more rational attitude can be attained to allow necessary humanitarian development in the future. It is axiomatic that any move must be within the limitations of what at the time is socially acceptable and has full regard for contemporary humanitarian and egalitarian principles.

With that in mind, I should suggest that the concept of the absolute value of human life may be progressively and acceptably relaxed in regard to the following.

(1) Abortion within the first ninety days of pregnancy should be available whenever there is a known risk greater than 10 per cent of serious abnormality in the infant, either from genetic causes or resulting from toxic or infectious processes involving the foetus *in utero*. In most advanced countries the law already allows abortion for such reasons, as well as where continuation of a pregnancy may have medically harmful effects on the mother. It is a highly controversial area, but the current trend in affluent non-Catholic countries is to allow abortion whenever it is asked for by the mother. In view of the threat everywhere of overpopulation, this use of abortion as a 'backup' for contraceptive methods must come eventually.

(2) When a child is born with a serious disability that can be recognized and assessed, preferably within a week, at the limit a month, after birth, that cannot be remedied by reliable surgical or

prosthetic procedures, and that would not allow the individual ever to occupy a satisfying place in the community or to have a normal sexual and family life, it should be subject to compassionate infanticide under appropriate legal sanction.

Here of course there will be strong opposition, but, as I have discussed in more detail elsewhere, such compassionate infanticide is already standard practice when the product of birth is such as to justify the term 'monstrous', i.e., where there is a gross and physically disgusting malformation such as anencephaly (complete absence of brain). Severe spina bifida, where there is no possibility of effective surgery, is also not infrequently dealt with by allowing the infant to die under sedation. Evenly balanced controversy persists in regard to spina bifida generally, the results of surgery being so unhappy that many paediatricians prefer to allow the child to die in comfort. One gathers too that the time is fast approaching when at least one genetic anomaly of metabolism (Tay-Sachs disease) will be handled similarly. Most physicians will agree that compassionate infanticide for this condition is no less morally defensible than the accepted routine in a suspect pregnancy of waiting three months until a cell test of foetal fluid (amniocentesis) can be carried out and, if positive, the foetus destroyed by a late abortion. In Tay-Sachs disease the infant is born blind and is grossly retarded mentally. Early death is inevitable, usually around the third year. In view of the significant dangers of amniocentesis to a normal foetus and of late abortion to the mother, there is every rational justification for waiting until a definitive decision can be made after birth as to whether the infant is normal or has Tay-Sachs disease, and acting accordingly. This is one of the situations where there can be no slightest doubt that compassionate killing of the diseased infant is the only humane approach, having regard to infant, parents, and the community.

Only when the position in regard to the three examples I have mentioned has been clearly stated, looked at, and accepted, will there be any possibility of this suggestion being thought of as a realistic approach to the reduction of one of the major sources of human misery. The mental trauma to the mother of compassionate killing of her infant will rightly be emphasized, but women have lost babies all through history and accepted it as the will of God. Nowadays an intelligent woman desires, and usually achieves, a two-child family; she would appreciate, rather than resent, anything that could help ensure that the two children she rears are genetically sound.

(3) When a person is diagnosed as suffering from a condition which, in the opinion of two or more competent physicians, will be

lethal with greater than ninety-per-cent probability within two years, the quality of the rest of his life should be clearly visualised for the patient so that he can consider the available alternatives. The typical example of such a situation arises when the patient is diagnosed as suffering from some form of cancer. Some types of malignant disease, e.g. the commoner types of skin cancer and early cancer of the neck of the uterus, can be treated with a high likelihood of a successful cure. Another situation is, however, also common, where an elderly man or woman finds himself or herself with a diagnosis of deepseated cancer of, for example, stomach, lung, or colon, which has already spread beyond the limits that would allow a reasonable likelihood of cure. In a young person, deepseated cancer is rare and when it occurs very likely to be fatal. Nevertheless it is a reasonable policy to urge that every possible means to eradicate the cancer should be attempted. In the old there is an alternative approach, which is to use simple symptomatic treatment with surgical or other major intervention only to provide comfort without any attempt to 'cure' the condition or to lengthen survival time. I believe that if such alternatives were carefully and honestly presented, most elderly people would opt for what comfort they can have rather than face mutilating surgery or other 'heroic' measures. In deepseated cancer the usual sequence begins with major surgery, often requiring secondary surgery or other measures to counter the changes rendered inevitable by extensive removal of tumour and adjacent tissues. This will normally be followed by X-irradiation and/or treatment with cytotoxic drugs; both involve destruction or damage to many normal cells and are liable to be associated with nausea, pain, and weakness. With survival for five years after visceral cancer around 50 per cent, I for one should prefer to live a shorter time than submit to what has become the 'expected' behaviour of both patient and his advisers. The circumstances of every individual case will be different, but what must be asked for is that the patient should be fully aware of the options and what in the way of pain, surgical accident, or misjudgement and so on he can expect according to the alternative chosen. If the patient chooses what the doctor regards as essentially passive euthanasia, he must be allowed his way.

In all these matters, the current need is to understand and describe clearly the present state of affairs between dying patient and doctor. It could well be found on analysis that the proportion of people accepting the view that their primary request is for prolongation of life under almost any circumstance falls steadily with increasing age. Eventually it could become an admired and even expected action that an old person should deliberately sign off from life when he realized that he had become a burden to his kinsfolk and the community. So it was with primitive people living on the margin of subsistence; in the difficulties of an overcrowded world it may again become the norm of admirable action, but we must not try to hurry the change.

(4) The final human situation that calls for reconsideration of attitudes to death concerns the man (it is hardly ever a woman) who is a danger to society. Men responsible for repeated violent crime, usually rape and murder, must logically be removed from society, either by life imprisonment or by execution. No other solution has ever been considered, but in a modern society the nature of execution for such conditions must not be 'capital punishment'. It is not punishment or revenge, but the orderly legal removal of a danger to the other members of the community. Psychopathic killers are rare enough for us to be certain that their abnormality is essentially genetic and therefore incurable. It would seem better even for the man himself that he should be killed without pain or public show rather than rot out his life in a prison asylum.

Everything I have discussed would be subject to abuse and injustice and error; movement in releasing the traditional taboo on civil killing must be slow and tentative so that abuses can be recognized and guarded against, but it must come eventually. Let us ask those who oppose it one question. This is a humanitarian programme designed to reduce the total of pain, misery, and indignity. If abuses should arise within it, will they be more than infinitesimal in comparison with the killing, torture, and intimidation that our male inheritance drives us to whenever political and religious differences split people into hostile in-groups and out-groups? Terrorism, civil war, national wars, world wars, genocide, and mass expulsions—we have had them all in my lifetime—and what may bring the end of us all is packed and ready for release in those hardened silos of the northern hemisphere. We have two terrible inheritances from our Pleistocene ancestors, and both involve an obsession with death. If we could succeed in rationalizing the first, it might provide a toe-hold for beginning to deal with the second before it destroys us.

8

Birth and the Burden of the Genes

Birth is the most blatantly mammalian of all human functions. It is painful, messy, and undignified, and when the infant emerges he greets life with a cry. He has come hither with his fate, barring accident, already foreordained, in outline at least, by the pattern of his genes. His parents must go on playing their biological role and the child must fit like all his contemporaries into the community and be modified to tolerate and be tolerated by it. Environment is all important, but most of the child's early environment will be determined by very similar genes to his own, made manifest in his parents' actions. Later he will learn to express himself in a wider human environment, but again the multitude of genes which govern community quality have a character essentially that of his own. Yet he is an individual in the sense that he is made and shaped by a unique set of genes never previously brought together in that combination and including a few which, thanks to recent mutation, have never before existed on earth. He may be fortunate in many of his genes, but it is inevitable that combinations of some will cause him to suffer. How much suffering will depend on the many random factors that determine which of the myriad possible combinations of genes that can be drawn from his two parents he has actually received. Everyone is aware of how widely children in a single family can differ from one another in almost every conceivable respect of appearance, intelligence, and temperament. In Chapter 2 an elementary account is given of how the male and female parental genes are, as it were, shuffled, split into half sets, and recombined in the new, fertilized ovum, to give again the standard (diploid) number of chromosomes. At this stage a little more detail needs to be provided in regard to the reduction division that prepares the reproductive cells for their function and the nature of recombination.

The sequence of actions in the reduction division (meiosis) differs as between male and female in some important features, but for our present purpose we can use a simplified outline of what happens which is applicable to the germ cells of both sexes. Each single set contains a large but still indeterminate number of structural genes, probably of the order of 10 000, and when the combination of two single sets to make the double set of the new individual takes place, 10 000 genes from the father arrange themselves in parallel with an equal number from the mother to make 10 000 pairs of corresponding genes. A large proportion of those pairs will be exactly similar, but very many will be made up of two genes differing in some significant fashion. The gene is represented, as we say, by two distinct alleles, and, in respect to that particular gene, the individual is heterozygous.

To go further in this discussion it will be simpler if, while still remembering that figure of 10 000 plus for the single (haploid) set of genes, we follow some diagrams in which the number is reduced to a much smaller number which can be represented by letters of the alphabet. In one individual we can have gene A, B, C, etc., each as a pair of two allelic genes. If they are both the same, no complications arise and we can represent the genetic makeup as AA or CC, and so on. Where alleles differ, however, they must be appropriately indicated in some such fashion as Aa or BB′. When a germ-line (reproductive) cell is ready for its reduction division, we can think of it as represented in Figure 5. In this example we are looking at nine allelic pairs of genes. Six of the pairs have identical structure; the individual is homozygous for those genes. In the other three pairs the alleles are different. One set of alleles comes from the father, one

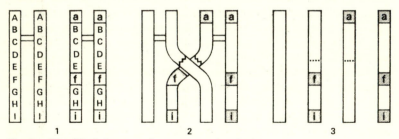

Fig. 5:   *Simplified diagram of cross-over at meiosis.*
1. Two corresponding chromosomes, showing three allelic genes, divide but remain as linked chromatids.
2. One chromatid of each pair breaks at an equivalent point and crosses over as shown.
3. The four resulting chromatids are distributed in the four haploid cells of the second meiotic division.

from the mother, and in due course a single chromosome of equivalent type will be provided for the sperm or ovum that will contribute to the next generation. But it will almost never be exactly like the two depicted on the first line in the figure. The new chromosome must have an A, B, C, . . . series of genes, but it can have any one of many different rearrangements. A certain degree of crossing over nearly always takes place, and one possible example is indicated in the figure. As the number and position of the cross overs in each chromosome pair is highly variable, it is evident that an infinite number of cominations of alleles of the A, B, C, . . . series of genes, and of every such series in other chromosomes, could be produced. Within the limits of the genes in a man's double set of chromosomes, each sperm he produces will be different from any other. Sometimes the difference between two sperms will be small in its effect on the eventual offspring, sometimes it will be large, and sometimes it will mean genetic catastrophe. Just the same holds for the differences between the 200 to 300 ova that develop during a woman's reproductive life. Each is in some way unique.

At fertilization one sperm meets and penetrates the single liberated ovum. Chance again comes into play to determine which ovum is to reach the uterus on the month of conception and which. individual sperm of the 300 million in the fateful ejaculate will win the race and fertilize the ovum. Once the two sets of chromosomes, of DNA, of genetic information have come together to form the zygote, most of the human-to-be's life has been sketched in outline. Accidents of many sorts may influence what he will become, but sex, intelligence, and constitutional health are already firmly foreshadowed and, barring accident, even his prospective lifespan could probably be forecast if we had an adequate knowledge and understanding of the infant's genes. I suspect that with full knowledge one could say with certainty that under the best of circumstances he will not live beyond X years and that if he avoids accident in all its significant forms he will almost certainly live to X − 5 years.

One of the essential objectives of this book has been to justify such a statement by analysing the influence of genetics on lifespan, but at this stage we need to go deeper. Inheritance, I believe, dominates every aspect of human life to an extent that few people are willing to accept. Fewer still would feel that anything was to be gained by examining the implications of such a conclusion for the future.

The range of human diversity

A newborn child may already have been deprived of some of his genetic potentiality for a healthy life. Even within the uterus the environment may not always be secure and peaceful. Viruses can enter from the mother's circulation, and particularly when she is infected with rubella (German measles) in the first three months of pregnancy very serious damage may take place—blindness (congenital cataract), deafness, and congenital heart disease being the significant results in the child. Drugs may also do harm, with thalidomide the notorious example. Most of the deformities of the newborn, such as harelip and cleft palate, club foot, spina bifida, and hydrocephalus, are probably in part genetic, with some environmental factor—none of them clearly defined as yet—also playing a part. Something over 1 per cent of births in Western countries are of a child physically deformed in some respect, and if we add most of the spontaneous miscarriages and stillbirths to all infants born alive with clinically significant abnormalities, the overall wastage is nearly 20 per cent of all conceptions.

Of those born apparently healthy there will be a small proportion who at some time in their lives will develop a genetically based condition giving rise to some degree of clinically recognizable disability or disease. The rest, the normal children, comprise those whose physique, appearance, and behaviour allow them to be included within the conventional definition of normality. It is these with which we shall first be concerned.

Genetics of the normal

Most of the physical characteristics by which we differentiate one individual from another are polygenic. At the structural level we tend to think of height, weight, skin colour, and facial configuration; and on the side of functional capacity, strength, athletic ability, and intelligence. All depend on the co-operation of many different genes. Their polygenic nature can be recognized by the characteristic distribution curve of differences when they are measured over large numbers of people from a socially homogeneous community, with due allowance for relevant factors such as age and sex. The so called normal Gaussian distribution of characters of this sort is probably known to everyone who has any serious interest in human diversity, whether of appearance or achievement. Anyone, in fact, whose concerns give him a reasonably wide association with men, women, or children will intuitively know the sort of result to be expected. If people are sorted into groups according to some numerical grading of the quality being examined, about half will be placed close to the average value. Outside this group the further the value is above or below the average, the smaller the group manifesting it. Adult males over 2 metres (7 ft) or under 1·25 metres (4 ft) high are extremely rare. If the results are expressed graphically, the distribution is shown as approximately to a bell-shaped curve that is often nearly symmetrical in shape. It does not necessarily follow that everyone who fits somewhere on a normal curve is normal in the sense of being a fully acceptable member of the community. For some qualities too much or too little may make a man an outcast.

Such statistical studies of the diversity of human form and function were initiated by Francis Galton, one of the great Victorians and in his own way almost as influential for the development of biological science as his cousin, Charles Darwin. If, as some critics may well hold, there is a somewhat mid-Victorian flavour about this book of mine, it is almost wholly due to the influence of Darwin and Galton. Each approved strongly of the other's work, and between them they seemed to look on the world with a combination of commonsense, intelligence, and imagination that attracts me enormously. Galton, I believe, laid down almost the whole foundation of human biology, and although he knew nothing of Mendel's work, he had an almost equal influence in bringing the science of genetics into being. His introduction of the use of twin studies to sort out the parts played in human development by heredity and environment, nature and nurture, has proved to be of special importance.

Twins occur about once in sixty-seven pregnancies, and of a hundred sets of twins, thirty to forty will be identical twins, the rest non-identical. The fact that some twins showed an extraordinary degree of resemblance while others were as dissimilar from one another as ordinary siblings had been known for centuries, but Galton was the first to grasp the biological significance of identical twins and to understand the essentials of their origin. We now know that they arise from what one might call a minor error in the first stage of development of the fertilized ovum. The primary nucleus divides, distributing its duplicated DNA in precisely equal fashion to each of the daughter nuclei. The rest of the cell, the cytoplasm, has also been redistributed and the two halves develop cell membranes to complete the separation into two daughter cells. Normally the cells should remain in contact and undergo a second divison at right angles to give four cells, and so onward, to build up the embryo. About one in two hundred times the two cells of the first division fall apart. Each still contains full genetic potentiality and in a highly 'biological' fashion each proceeds to develop into a separate complete embryo. A pair of identical twins results. They are genetically identical, of the same sex and blood groups, and strikingly similar in appearance. Non-identical twins appear primarily as a result of the liberation of two mature ova at the time of ovulation instead of the usual one. Each has been fertilized by a separate spermatozoon, and when they implant in the uterus they develop as two normal embryos with a minimum of interaction. Except on rare occasions, the two placentas develop separately with no fusion of the two circulations. Such twins may be of the same or different sexes and resemble each other no more than other siblings in the same family.

The twin study technique can be exemplified by a famous investigation planned in the earlier years of the twentieth century to decide the relative parts played by infection and heredity in the development of tuberculosis. The principles by which the twin study method is applied to any type of clinical problem have remained the same ever since. In this instance the investigators had access to the case records of large numbers of persons with active tuberculosis of the lungs. Among these, as for any random human population, between 1 and 2 per cent of the subjects are likely to be twins, and for most of them a co-twin or a deceased twin's medical record can be located. The investigation was set in motion, co-twins were found, and family studies begun. Once brought together, each pair of twins was examined to determine whether they were identical or non-identical and whether the co-twin had or had not active tuberculosis or other evidence of infection by the tubercle bacillus. In practice there were, and are, many difficulties to be overcome in such studies, but the principle is perfectly simple. If the concordance of identical twin pairs is significantly higher than that of non-identical pairs, there is an inherited element in the condition. In their famous paper, written so many years ago, Kallman and Reisner[1] found that 87 per cent of identical twins were concordant in that both had suffered with clinical tuberculosis. Non-identical twins showed only 26 per cent of co-twins with tuberculosis, exactly the same percentage as was found in other brothers and sisters of the index twins. Susceptibility to tuberculosis is therefore almost as dependent on one's inheritance as on infectious contact with the tubercle bacillus.

So far we have been dealing with conditions resulting from the interaction of many genes and we have been concerned with the distribution of normal qualities, including resistance to infectious disease. The other group of significant differences among people are those that show what we call Mendelian inheritance. Like the differences between the races of garden peas that Mendel studied, these concern conditions where change in a *single gene* produces the differences that we can observe. Sometimes these differences involve normal characteristics, but in man most of the easily recognized examples of Mendelian inheritance concern genes producing abnormality or disease. Irrespective of this, when we deal with any

Mendelian condition we find nothing that resembles the bell-shaped Gaussian curve but a clearly discontinuous distribution. As examples of normal qualities, all people are either male or female—sex is a genetic quality—and when the ABO blood group is studied, everyone is AB, A, B, or O. There are many other biochemical differences found in healthy persons which also show a Mendelian distribution, but they are of little practical importance. The clinical study of human genetics tends to be concentrated on gross genetic disease, where a Mendelian inheritance is either directly demonstrable or can be deduced with considerable certainty by indirect evidence.

The types of genetic abnormality in man

A host of medically significant abnormalities, some relatively common, many excessively rare, are genetically transmitted in true Mendelian fashion. They can be divided into dominant, recessive, and sex-linked, and the mode of inheritance is shown in Table 3. This is the first thing to be taught in elementary genetics and in some ways it has been a hindrance rather than a help to the development of the science, but it still has its uses. The sex-linked quality depends on the fact that the gene responsible for the abnormal quality is carried on the X chromosome. Such conditions are recessive in the female, where there is a normal X chromosome to dominate the abnormal gene, but dominant in the male, since he has only one X chromosome. The rules of inheritance are straightforward if we remember that, except for X and Y sex chromosomes, every gene is associated with an equivalent gene in the other member of its

Table 3

DOMINANT	Aa × AA → 2AA + 2aA		+ 50%
	+ − − −		
RECESSIVE	Aa × Aa → AA + 2aA + aa		25%
	− − − − +		
	Aa × AA → 2AA + 2aA		0
	− − − −		
SEX LINKED	XY × Xx → XY + xY + XX + xX		M 50%
	− − − + − −		F 0
	xY × Xx → XY + xY + Xx + xx		M 50%
	+ − − + − +		F 50%

Table 3:   *The three types of Mendelian inheritance.*

chromosome pair. If the genes differ, they are spoken of as alleles, and in most Mendelian conditions one allele is normal and the other disadvantageous. We are interested of course in the disease-provoking gene and speak of it as dominant if whenever the gene is present in the genome in the presence of a normal allele the abnormality is expressed. As can be seen in the standard formulation, when a person with a genetically dominant abnormal character mates with a normal individual, one half of their children can be expected to show the abnormality. When the abnormal gene is recessive, it will in practice only appear when two apparently normal people, each with one abnormal and one normal gene, have children. On the average, one-quarter of the children will show the abnormality. There are a number of conditions, mostly concerned with relatively minor chemical differences between the products of two alleles, when both forms will be expressed simply according to their presence in the genome. These are spoken of as co-dominant.

When the sex chromosomes are concerned, the situation is only a little more difficult to follow. The sex of an individual is determined in the first instance by the constitution of the pair of sex chromosomes; if it is XX, the individual is female; if XY, male. If one allele of a gene carried on the X chromosome is abnormal, that abnormality will be expressed in every male who receives the faulty allele. So when a woman who carries the gene without symptoms marries a normal man, half of her sons, but none of her daughters, will show the abnormality. This pattern is spoken of as sex-linked inheritance.

An important technical use of sex-linked characters becomes possible when two alleles of a gene carried on the X chromosome are co-dominant. They can be called A and B. For some reason the Y chromosome carries very few genes, the X chromosome a great many. Perhaps to bring the male and female genomes into equivalence, one of the two X chromosomes in all female cells must be rendered inactive. This is accomplished by a unique process, the Mary Lyon phenomenon, initiated early in embryonic life when the embryo contains only a few thousand cells. In each cell one X chromosome closes down and plays no active role for the rest of the individual's life; all the activity needed by that cell and its eventual descendants is supplied by the remaining X chromosome. Which chromosome, A or B, closes down is apparently determined wholly at random—by tossing pennies, as it were. About half the cells express A, and other half B; none expresses both. As the embryo develops into a woman, descendant cells multiply, migrate, and give rise to the organs of the body. Each organ turns out to be a mosaic in which little groups of A cells are distributed among similar groups of B cells.

This allows a general method of determining whether a certain tumour or some other type of cellular proliferation is monoclonal or not. The importance of this in relation to somatic mutation has been discussed on page 66. The actual quality expressed by A and B in the only commonly used method is the antigenic quality of a certain enzyme, always referred to as G6PD. There is no need to explain any more about it than to say that the forms A and B are quite common in people of African origin, and among any score of American black women at least one carrying both alleles will be found. There are convenient methods of testing for A and B in the laboratory, and very many of the common tumours have now been tested. Almost all have proved to be monoclonal.

Genetic pathology

To return to less technical matters, we are concerned primarily with the advantages and disadvantages—often crippling disabilities—that the infant brings in his genes at birth. To any thoughtful physician a genetically handicapped child is a subject for compassion but also a challenge to his scientific knowledge and understanding. Every week or two some recently born infant is found to be suffering from a genetic mishap that seems to differ from any that has previously been described. If the child is found in or sent to a university teaching hospital, he will almost certainly become the object of sophisticated, long continued and expensive research to find the function of the gene that is responsible. It is part of the traditional wisdom of medical research that 'experiments of nature', genetic abnormalities of human beings, may often tell us more than all our physiological experiments on normal laboratory animals. Part of the research will be a search for ways of ameliorating or curing the defect by relevant methods, sometimes surgical, sometimes by artificial prosthetic aids like spectacles or splints, sometimes by biochemical manipulations.

There is an infinite variety of genetically based deviation from the normal. Sometimes the error in DNA structure has so damaging an effect that the embryo cannot survive and dies at some stage before birth. Probably a majority of miscarriages are genetically caused. Among the congenitally abnormal infants born alive there are very few specific types of genetic disease that occur as frequently as one in a thousand births. Many are very much rarer, but if we include both mild and severe types the overall number becomes very large.

If one takes account of all the deformities and clinically significant anomalies with which a child may be born, the number approaches 5 per cent of all births. Many are not wholly genetic, but there is a genetic component in virtually all of them. Much has been written about genetic disease, and there is a standard catalogue of those that are transmitted in Mendelian fashion by McKusick[2] of Johns Hopkins Hospital. He lists 415 certain and another 528 less adequately studied types of dominant genetic disease in man. The figures for recessive conditions are 365 certain plus 418 possible; for sex-linked conditions, 86 certain and 64 possible. Overall this adds up to a total of 1876 named disabilities of diseases and a new one is described every month. The number of disabling conditions of polygenic origin is unknown. From their nature they are difficult to define; each type must almost necessarily grade from the most outspoken form continuously into the normal range. Most types of mental retardation and many forms of psychosis probably belong here.

Even without adding a figure for polygenic genetic disease, the rough and incomplete figures I have given add up to an appalling total of human misery for the children, of intense unhappiness for the parents, and of expense to the community for their care.

It may be helpful at this point to give a brief sketch of four fairly typical and well known genetic diseases: two recessive Mendelian conditions associated with biochemical abnormality (phenylketonuria and Tay-Sachs disease), a combined genetic and environmental malformation (spina bifida), and a chromosomal abnormality (Down's syndrome or mongolism).

Phenylketonuria (PKU) is the only genetic disease that is widely tested for in babies at birth and whose effects can be mitigated in a considerable proportion of those recognized. The name refers to the fact that children with the disease excrete abnormal metabolites, 'phenylketones', in the urine; this can be recognized by a simple colour test. The basic abnormality in most cases is the failure of the mechanism to transform one amino acid, phenylalanine, into another, tyrosine. This leads to a dangerously high concentration of phenylalanine in the blood whose main effect is to produce by some action on nerve cells a severe degree of mental retardation and some deficiency in pigmentation. Phenylalanine is an essential element of the diet, but if such children are fed on semisynthetic diets low in phenylalanine, the concentration of the substance in the blood can be kept below the danger level. There is no doubt that children on that regimen on the whole do better than those untreated, but the results have not been wholly good. As is the way of a genetic disease when it is investigated in detail, new complications begin to be recognized that may interfere with both diagnosis and treatment.

Several genetic processes can give the positive PKU test and only some are associated with high phenylalanine in the blood. It is impossible in any given case to know what the intelligence would be if the metabolic defect were completely righted. Nevertheless, the most recent reports that I have seen claim that infants who were recognized as affected with typical PKU immediately after birth and given the appropriate diet have at ages five to seven shown the normal range of intelligence as judged by standard tests. It still remains to be established that the children will grow up to adulthood with normal intelligence, but those concerned are very hopeful. Difficulties with some children are bound to occur, and since phenylalanine is an 'essential amino acid', one must tread a metabolic tightrope in treating these cases; too little as well as too much can cause severe damage.

Tay-Sachs disease is another autosomal recessive metabolic disease in which an abnormal fatty substance is progressively deposited in nerve cells. This produces gross mental retardation, blindness, and behavioural disorders. Death is invariable by the age of three or four years, and it is unlikely that any effective treatment could ever be devised. Of all genetic diseases, this is the one for which it is most obvious that the only rational treatment is quiet infanticide. Since this is legally murder, the only practical approach is to test foetal cells from the amniotic fluid of any pregnancy in a marriage where the parents are known or suspected of carrying the gene, and abortion of any foetus shown to have the metabolic anomaly.

With spina bifida we come to an entirely different group of genetic diseases. In a typical case the child is born with the lower end of the spinal cord open to the environment and a variety of secondary anatomical changes in the same general region. It is certain that some nongenetic factors are concerned. The incidence varies from year to year; around 1930 the incidence in America rose to about three times what it was in the years before 1920 or after 1945. The nature of the environmental factor responsible for this unrecognized epidemic is unknown. A case against something associated with disease in potatoes was presented a few years ago, but it does not seem to have held up under criticism. A genetic factor is generally accepted as playing a part, but the exact form of this is also obscure. The only treatment available is surgical and the end results have been so unsatisfactory that at the present time there is a growing consensus that surgery should not be attempted and the short life of the infant made as tolerable as circumstances allow by drug sedation and careful nursing. In blunter terms, this is quiet infanticide.

The fourth example is the best known of genetic diseases, once known as mongolism, now as Down's syndrome, or by medical geneticists as 21-trisomy. It is the prototype of a chromosomal disease, in being regularly associated with three instead of the normal two examples of the small chromosome No. 21. As is well known, the disease is commonest in children born of mothers approaching the end of their reproductive period; it may be as high as 1 in 60 for mothers over forty five as compared with an overall frequency of about 1 in 600. Many affected children die in their first year. Those that survive grow up with a severe degree of intellectual weakness which usually precludes any possibility of self-support. It is notable, however, that they are usually happy-natured, affectionate, and readily handled. Throughout life they are unduly subject to infection, and life expectancy is considerably lower than normal.

All genetic pathology is the pathology of error, but the form of error is quite distinct, in Down's syndrome, from a recessive condition like PKU. The error is made during the reducing division which gets rid of half the maternal chromosomes. Perhaps due to some change associated with the approach of the menopause, one chromosome 21 fails to move as it should to the group that will be discarded, leaving the mature unfertilized ovum with one too many chromosomes. With the advent of the fertilizing sperm, the total becomes forty-seven instead of forty-six owing to the presence of three examples of chromosome 21.

Many other examples of chromosomal abnormality are known, but the only point that is relevant here is that most of these abnormalities, and in particular trisomy of chromosomes other than No. 21, are usually lethal to the embryo. Around 35 per cent of spontaneous abortions have been found to show chromosomal anomalies.

With the somewhat uncertain exception of PKU there is little that is useful to be done about any of these conditions. I have found nothing about the expectation of life of treated and untreated PKU babies. The other three conditions all lead to early death, and when death comes, the only comment that is possible is surely 'It were better that he had never been born'.

The prevention of genetic and other types of congenital disease

Nature has never had any mercy on the runt or the weakling. Gregarious animals will often kill or expel from the herd any individual of abnormal appearance or behaviour. In Africa the young of any of the hoofed mammals are subject to attack by predators almost from the moment of birth. Any newborn animal with less than normal ability to recognize danger and flee from it is doomed. Evolution, like the breeder of domestic animals, has only one way of dealing with genetic abnormality that has no promise of leading to better things. Any progeny that are below standard are killed; there is never any difficulty in breeding more—many more.

Man being what he is, nature's approach has become inadmissible. For centuries men have had to struggle with the ethical significance of a whole series of biological problems that they are only now beginning to understand. Not all of them are genetic. Overpopulation is probably the most important of all, deterioration of the biosphere the one that provokes most public attention, and nuclear war the most sinister threat. Nevertheless, in the very long term genetic deterioration will become the outstanding problem for those who want to see an indefinite continuation of civilized man.

In the affluent countries, from which comprehensive demographic and medical statistics are available, there is a heavy burden of genetic disability, as I have described in previous sections of this chapter. Indirect evidence and all biological logic indicates that the genetic quality of the species, as measured by physical vigour, absence of disease of intrinsic origin, and intelligence, is slowly deteriorating, so slowly as to be hardly detectable in a generation, but from the evolutionary angle at a positively disastrous rate. In the absence of the primitive winnowing process of natural selection, an inexorable increase in the burden of faulty genes must be going on in almost all human populations. If we are to have any forethought for the long-term future of the human species, some human initiatives to counter that process will be needed. The second portion of this chapter represents an attempt to discuss the possibilities of interfering successfully with the mammalian reproductive process to that end.

Too many people

In a world where men and women could act rationally in dealing with the problems of overpopulation and genetic deterioration, the need would be quite clear, first, to reduce the birth rate to a level that would allow the population to fall to a number appropriate to the world's resources and then stabilize it with an average of about 2·3 children per completed family, and, second, to ensure that all surviving children should be within accepted bounds of genetic normality. The two logical solutions must, however, be looked at separately. As yet, birth control is not really considered in relation to the incidence and increase of genetic abnormality. Planning for a family of two children may be supported by demographers and philosophers as a means of saving the world from overpopulation, but 99 per cent of those who actually practise birth control do so because they want to avoid the relative poverty and extra effort that a larger family entails and to have the opportunity as soon as possible to bring two pay packets instead of one into the household.

Birth control was a reality long before the advent of the contraceptive pill. Methods, however, were crude and aesthetically unattractive and they were never discussed in the public press. The Pill has changed all that. It is such a simple business to take a pill every morning for twenty-one days of the lunar month and enjoy one's husband in comfort and without fear of consequences. Contraception became a good thing, easy to talk about, and acceptable matter for discussion in print almost anywhere. Even the Roman Catholic hierarchy were at least divided in their opinion about birth control, and for that part of the world that knew about it and could afford it, it was an unalloyed boon that must have done more for human happiness than any other achievement of science. No one was pedantically logical enough to point out that once a month two living cells, an ovum and a sperm, which might have become a human being were being casually murdered.

Nor was there any suggestion that methods aiming to prevent the implantation of a fertilized ovum by a 'morning-after' pill would be any less acceptable. In talking about these matters I have several times confessed to a certain worry about using slightly unnatural hormones to interfere with the normal sequences for twenty years or more. I felt that ultimately the ideal method could be an improved safe-period formula with a drug (one of the prostaglandins, perhaps) to induce a very early abortion or activate a delayed menstrual period by administering it as soon as a period was significantly overdue. In all of this the significant feature is that nobody concerned with any of these methods would have any knowledge or intuitive sense that potential human life was being destroyed.

Abortion before the third month of pregnancy if the child for any reason is unwanted is very widely accepted as a fundamental right of any woman, single or married. At the same time the equally important new dictum that a single woman has as much right to have a baby as a married one, if she wants to, is nearly but not quite so universally accepted. Part of the acceptance of abortion as a routine solution for unwanted pregnancy probably relates to the fact that in the course of an abortion the woman never sees the foetus; it is very much like having a tooth out at the dentist's, and psychological trauma can be minimal. Infanticide remains unthinkable. Once a woman has seen and handled the baby, an intense bond develops rapidly, and it is rightly felt that to sever that bond by eliminating the infant would be unbearably cruel to the mother.

This, then, is approximately the position at the present time of the prevention of pregnancy in quest of an easier life or to help reduce the degree of overpopulation. As far as one can gather, in affluent societies birth control is practised by the majority of people in all social classes and is not much influenced by their nominal religions. Contraceptive techniques are relatively effective, especially if backed up by abortion, and there are interesting possibilities of evolutionary changes being brought about by these types of interference with reproduction. In the field of genetic disease and disability, however, birth control has as yet no more than minimal relevance.

Diminishing the impact of genetic abnormality

The first approach to the problem of genetic deterioration is a short-term one which is not likely, even if it were generally adopted, to greatly diminish the proportion of genetically abnormal infants to be born. The action suggested is the second phase of what was called the only logical solution: to make sure that all children reared were within the accepted limits of normality. To put it bluntly, this means killing the product of conception as soon as its inadequacy to face life is known with certainty. This of course may be acceptable enough when the process of killing can be masked under the guise of delayed contraception or abortion, but it is another matter if the killing takes place after birth at term. This may be spoken of as mercy killing or compassionate infanticide, but legally it is murder. However, in my own mind I am confident that within less than a hundred years such action will be accepted as socially necessary, morally acceptable, and perhaps even compulsory under law.

Something of the sort is already happening, as I have hinted earlier when speaking of spina bifida, and a little more may be said here about another example of the same group of diseases. It is known as anencephalus, in which the infant is born a repulsive-looking monster with virtually no brain development and only a flabby bag behind the face. When I was taught obstetrics as a medical student, we were told that if such an abomination was born alive, the proper action was to be sure that the mother did not see it and to give it an injection of morphia so that there could be no pain while it was being allowed to die. Spina bifida is an essentially similar process, involving the lower part of the spinal cord, and, as I have said paediatricians are about equally divided between allowing a child with severe deformity to die under sedative drugs and attempting surgery. The great majority of parents will accept the doctor's advice whichever way it goes, probably because in an open spina bifida, as in anencephalus, there is a grossly visible deformity. This is enough to balance, or more than balance, instinctive affection toward an infant and the pressure of community sentiment against anything savouring of infanticide. There are some other 'monsters' which are also subject to infanticide by universal consent: grossly incomplete 'Siamese twins', cyclops (one eye only), etc.

My point is, of course, that when the inability of a product of conception ever to have a tolerable existence is utterly evident to the eye of anyone, infanticide, either positively or by neglect, is already acceptable. The changing approach to spina bifida indicates too that among both obstetricians and the community such use of quiet infanticide is widening its range. There is a very slow release under way from that taboo.

At the present time there are serious efforts being made to avoid the taboo on infanticide by finding a way to recognize the existence of anencephalus, spina bifida, or related conditions, by some test that will allow the condition to be detected early enough to allow an abortion. Except in traditional religious circles, the taboo against abortion has almost vanished and a sharp distinction is recognized by most people between abortion, which is killing the foetus by expulsion from the uterus before it is viable, and infanticide, which is killing the product of conception when it leaves the uterus in a viable state. It is recognized in the conditions I am referring to that in neither case is it possible for the product of conception ever to have a tolerable life. Unfortunately, the test for α foeto-protein in the mother's circulating blood, although positive in 80 per cent of cases in the second half of pregnancy, is never diagnostic during the first three months that are available for safe therapeutic abortion.

There are of course many other genetic and congenital malformations that can show every degree of severity and of amenability to surgical repair. Mild harelip, pyloric stenosis, and imperforate anus are examples of serious or potentially fatal malformations that are almost regularly remedied completely by any competent surgeon; but there are many intermediate between these and open spina bifida.

But in additon there are at least a thousand serious genetic abnormalities where the infant is born anatomically normal. I have spoken about the two most gross anatomical malformations, and earlier in this chapter I have described Tay-Sachs disease as one of the most distressing of genetic metabolic diseases. This and another even more damaging abnormality of infant metabolism, Lesch-Nyhan disease, are both recessive conditions in which a family with one affected child can expect that one out of four children subsequently born to them will have the disease; the other three will show no overt signs. By the use of refined biochemical techniques it is possible in principal to recognize that a foetus is one of the unfortunate 25 per cent by testing cells from the amniotic fluid at about three months' gestation. In a few instances, I believe that positive cells have been obtained by this method of amniocentesis as it is called, and an abortion performed.

The problem of the congenitally abnormal child is perhaps the most prolonged and difficult of all those that can confront parents under today's conditions. With the only 'logical solution' inadmissible, a wide range of treatment has been suggested. For technical reasons that need not be detailed, there is not the slightest likelihood that any method will be developed to eliminate genetic weaknesses in the afflicted individual. This is not to belittle what can be, and is, done to make life more tolerable for the sufferer. The correction of minor visual defects, which are probably the commonest of all genetic deficiencies, by appropriate spectacles is the prototype of much that can be done by prostheses, plastic surgery, special diets, and so forth. What must be realized, though, is that the existence of a life-long genetic cripple, if he is to be adequately handled by modern methods, requires almost the full-time service on the average of at least one professional. Two people are prevented from undertaking useful work for each such handicapped individual, one being the victim and the other his proportional share of the army of medical and paramedical professionals and their aides that are involved. Again we have to recognize the futility of keeping genetic cripples alive—meaningless lifetime misery for the victim, endless worry and social difficulties for the parents, and heavy, unprofitable expenditure by the state.

Eugenics

Eugenic possibilities are not popular topics for discussion in the 1970s, perhaps for the wrong reasons. Eugenics, whether theoretical or practical, must have the working assumption that some human beings are better than others—are more desirable components of future human populations than others. In Australia, and probably in most other effectively democratic countries, we have a contempt for tall poppies in any field where the ordinary person feels he or she could never hope to excel. Senior administrators, successful businessmen, writers, and scientists must walk warily or they will be cut down to size; musicians, especially pop musicians, and all types of athletes and those who excel in sport are free from that inhibition. They are not suspect of being élitist. Anyone who can think hard and effectively about any topic is placed in the élite, and the average man is rightly suspicious that those who think about and advocate eugenic measures are concerned to increase the proportion of the élite as so defined in the community. His other article of faith is that no person should be disadvantaged for something that is not his fault: neither physical crippling nor genetic defect should in his view debar anyone from the satisfactions of marriage and parenthood.

None of this is relevant to the discussion except in its bearing on political attitudes to demographic policies.

In the present state of knowledge of human genetics and of public attitudes to interference with privacy by the state, there is no place for anything connected with eugenics apart from very long-term discussion of what may eventually be inescapable. In almost all respects a long-term regard for the genetic quality of the human species has the same status and calls for analogous types of action as the accepted responsibility for maintaining the quality of the environment. We do not even dream of planning a perfect environment, but we accept a responsibility to monitor the changes that are constantly occurring insofar as they seem relevant to human health and well being. Air and water pollution takes first priority, but serious distortion of ecological situations or destruction of natural beauty are becoming of almost equal importance. Where we are concerned with human populations it is routine and indispensable that there should be constant monitoring of numbers, ages, and distribution of people in the administratively defined community. Occupation, salary or wages, other financial resources, type of dwelling occupied are all on record somewhere in official files. Monitoring of health and physical capacity for work is to a considerable extent in the hands of the individual, but the state becomes more deeply concerned when the problem concerns infectious disease. In this field, control becomes more and more authoritarian with increasing danger, rarity, and public fear of the disease. Cases of smallpox, bubonic plague, or cholera are treated without regard to privacy or the rights of the individual. This holds also, but in a different fashion, for leprosy. In the related field of veterinary quarantine, the appearance of cases of rinderpest or foot-and-mouth disease in Australia is followed immediately by the slaughter of all healthy, susceptible stock within a radius of as many miles as seemed called for by the epidemiological circumstances. Whenever the risk to the life or economic well-being of other people is great enough and immediate enough, individual rights can be completely overridden by the state. In the development of this approach most of the actions called for by any eugenics policy have been adumbrated in principle. They will, however, never be used until people have a sense that the danger they are designed to forestall is great and immediate.

In the interest of monitoring the position, the important need is to have active academic work on human genetics in universities and research institutes and a good working relationship with government departments of health, social welfare, and demography. We must be able to follow the long-term trends of such genetic diseases and congenital conditions of partial genetic origin as mongolism, spina bifida, cystic fibrosis, haemophilia, and PKU. Others will be added as knowledge increases. Such survey work will be inevitably associated with research on many possibly relevant factors that are concerned with the spread of the peccant genes and with the regularity and degree to which the genes are expressed in those who carry them. One of the significant features of modern medicine is the recognition that every significant ailment that brings a person for medical advice is dependent on many causes.

I have come to believe that in every chronic degenerative disease in the last third of life—from fifty onwards—one can usually distinguish:

(1) genetic factors;

(2) environmental influences, often the triggering effect of a relatively minor infection, not infrequently a drug used over a long period;

(3) genetic error in somatic cells. The nature and some of the significance of somatic mutation have been dealt with in earlier chapters and will be raised again in regard to age-associated disease. Here it need only be noted that three types of effects of somatic genetic error are of clinical importance: (a) cancer and neoplasms in general, (b) autoimmune disease, (c) manifestations of normal ageing;

(4) psychosomatic influences which in their turn will probably be found to be largely genetic in nature.

Even that probably falls far short of the complex reality of a patient's disability, but at this point it serves merely to show the continuing need for research on the interaction of genetic and non-genetic factors in causing disease. I believe that in the long term the only legitimate objectives of any policy that could be labelled eugenic will be to reduce the number of children born with clinically definable disabilities that are not amenable to socially and economically preacticable forms of effective treatment and that make a tolerable life impossible. Such an approach is in fact being actively followed in many parts of the world and has given rise to a new type of medical specialist, the genetic counsellor.

The practical problems on which such a counsellor's advice is sought will take many forms. The commonest is whether, after a child with some congenital abnormality is born, the parents should have any further children; and probably the next most frequent is for advice on a projected marriage in which one of the partners has one or more close relatives with a presumed genetic disability. For most such problems the answer can only be given in terms of such and such a probability. To the first question the most usual will be to say that the likelihood of another child being similarly affected is around one in four, with the qualification that the degree of disability manifested in an affected sibling may be more or less according to what is spoken of as the *penetrance* of the abnormal gene. That could be a completely honest answer about the disease cystic fibrosis for example, the commonest of the recessive genetic diseases, but obviously it is far from giving a specific basis for deciding whether or not to have one more child. Questions on the advisability of A and B marrying and starting a family will usually be even more difficult. Here, however, there is optimism in the air. Already there are some recessive conditions (in which, it will be remembered, both genes of the pair must be present to give symptoms) in which it is possible to test a person without symptoms and say whether he or she is a heterozygote with one abnormal gene or is wholly unaffected, with two normal genes. Once this is known, it becomes possible to say in most instances that as far as the condition they are worried about is concerned, there is no risk of its appearing in any children of the marriage. In the unlikely occurrence of two people heterozygous for the same harmful gene wishing to marry, the counsellor will probably advise that each should seek another partner.

If this possibility of detecting symptomless heterozygotes with certainty is extended to most of the significant recessive conditions and the dilemma of two heterozygotes wishing to marry and have children becomes relatively common, the logic of allowing quiet infanticide may become more insistent. In a world where two children is the norm, many couples would be happy to take the 'even money' chance that *their* two children would both be normal—provided that if an affected child were born it should be quietly eliminated and replaced at a subsequent pregnancy. Where the existence and intensity of a risk is definitely known, it is only human nature to tolerate any logical way by which that risk can be circumvented. It may not be too many generations away before all infants known to be at risk will be tested for those genetic diseases that can be recognized within forty-eight hours of birth and returned to the mother only if all is well. The technical requirements for such testing on a large scale are rather daunting, and it seems possible that there may even be a public demand for such a service before it is practical to supply it.

One can sympathize with a woman who says in a hundred years' time, 'My husband and I would be happy to have a family of four or six—the state and public opinion say that I can have only two living children; if I accept that, it is surely for the state to ensure that both those living children of mine are free from any genetic fault that will grievously harm a life.'

All that I have so far written under the heading of eugenics has really no other objective than to show compassion to those individuals predestined to the tragedy of intolerable life. It is not directed toward any idea of producing a better human species. Not until there is a change in human values, in public opinion, and in the prejudices of rulers, will it be even possible to consider a true eugenic approach. It is just not practical politics, and I shall only mention it again in some final speculations about the distant future of man's descendant species.

9

Xeroderma Pigmentosum

Around 4 per cent of all the infants that are born alive are in some way congenitally abnormal. In a proportion, the abnormality will be evident at or soon after birth, but there are some inborn genetic anomalies that will not be evident until middle age. As I have discussed in Chapter 8, there is an immense variety of such defects, many of them relatively trivial or easily remedied by surgery. By no means all of them are known to be genetic in origin. Some are due to infection of the foetus in the uterus; others, such as the diminutive limbs of thalidomide babies, result from exposure to toxic substances during the early stages of pregnancy. Most, however, are either of direct genetic causation or have a significant genetic component. This chapter deals with a rare genetic disease, an autosomal recessive condition, in which the most obvious abnormality is in the skin that during childhood becomes dry and densely covered with freckles—hence the name xeroderma pigmentosum, which henceforth will be referred to, as it is in all technical talk about the disease, as X.P.

To devote a chapter to a disease that very few doctors have ever seen and that occurs only once in every quarter of a million births requires some justification, and it is probably best to start with a few general remarks about the significance of genetic abnormalities for the understanding of how the body functions in health as well as in disease.

Experiments of nature

Particularly for a genetically determined condition like X.P., there may well be good reasons why a disease can never be prevented or effectively treated. That, however, can never be predicted until the nature of the disease is understood. Some of those who are concerned only with the primary requirement of medicine, that the patient should be cured of his illness and rehabilitated into his place in the community, can on occasion be critical of the way clinical research in our teaching hospitals tends to concentrate on under-standing disease processes rather than on improving treatment. I can sympathize with some of these objections, but must still hold firmly to the importance of understanding what is happening in disease and accept this as the primary objective of clinical research. Such work calls for curiosity, intelligence, and imagination on the part of the investigator, and it has never been easy.

The normal functioning of the body is something that most people—and their physicians—can take completely for granted. We have learned to accept that 'we do what we want to', without any necessary knowledge of the nervous and muscular mechanisms that are brought into action; and many other bodily functions are even more effectively hidden from consciousness. One might almost say that nature has taken a pride in concealing the fantastic complexity of its mechanism. Often it is only when certain types of genetic change occur that we obtain insight into just how some special function is carried out. As an example I can mention the discovery soon after the advent of antibiotics that a very small proportion of infant boys were born intensely susceptible to pneumonia. Before penicillin, the pneumonia was always fatal; with careful treatment with antibiotics the children could recover, but remained just as susceptible to any other bacterial infection of the lungs. Something was clearly wrong with the immune system, and Bruton in 1952 discovered that one of these children could produce no antibody and almost completely lacked the particular protein of which antibodies are composed, then called gammaglobulin. In due course that dis-ease, still called agammaglobulinaemia, was shown to be of genetic origin, a sex-linked condition affecting only males. Happily, it was amenable to treatment; by giving monthly injections of concen-trated globulin from normal people and taking special care of any episodes of infection, many of these children can be given a reason-ably healthy life. Once this had been established, another very in-teresting and unexpected discovery was made. Inevitably some of the children were accidentally exposed to measles—and, surpris-ingly, their attack of measles was a perfectly normal one—and if they were protected against subsequent bacterial infection by antibiotics, recovery was rapid and uncomplicated. Even more unexpectedly, it soon became clear that, once having had measles, these agamma-globulinaemic children were just as immune against a second attack as normal children, even though they produced no antibody against the measles virus. For several years the combination of findings was inexplicable to immunologists, but in the last decade it has fallen into line with advances in experimental immunology and supplied essential information about the respective roles of what we now call T and B lymphocytes. In a real sense such genetic diseases are 'experiments of nature', using techniques that are not available in the laboratory and that are carried out on the 'animal' we are primarily interested in, the human being. X.P., in my view, is the most illuminating of all nature's experiments.

The picture of X.P.

X.P. being a recessive disease and extremely rare, the parents of the affected child show no symptoms, although we can be certain that both carry one of the abnormal genes. So do 0·4 per cent of the population, but the chance of two such people marrying is 1 in 62 500 (250^2), and the probability of their children being affected is, according to standard Mendelian rules, 1 in 4. Hence, in round figures, the occurrence of about one case in 250 000 births.

The infant with X.P. is usually born without visible abnormality and for a few months develops like a normal infant. The first sign that something is wrong appears when the child is first taken out of doors. Any skin areas exposed to sunlight become acutely inflamed and blistered. It is immediately obvious that the child is quite abnormally susceptible to sunburn and from then onward he is protected as far as is practicable against exposure. Even so, some impact of indirect sunlight is unavoidable and this is evidenced by low-grade skin inflammation and the appearance of many pigmented spots, freckles of a variety of shapes and sizes, on all exposed areas of skin. The eyes too suffer from inflammatory changes. It is an unhappy life for the patient, but worse is to come.

About the time he enters his teens, the X.P. child is likely to develop his first skin cancer. By this time all exposed areas are thickly spotted with freckles and small moles. In between the spots, the skin is dry and atrophic, like that of an old man, and here and there can be seen small, scaly accumulations of cells. The first cancer is evidenced by the appearance of a hard, spreading lump which soon develops an ulcerated centre. It is essentially the same as the common rodent ulcer (basal-cell carcinoma) of old people, and, if treated early by surgical excision, is readily eliminated. From then on, however, the cancers continue to appear, and in former days almost all patients died of cancer before they were twenty-five. Under continuous medical observation, with excision of tumours as they appear, life can be almost indefinitely preserved. Several affected persons have had more than 100 tumours sugically removed and proved under the microscope to be true cancers. Most of them are cancers of the epidermal cells and relatively easy to handle, but occasionally a more sinister tumour arises from pigmented cells of mole or freckle. This is malignant melanoma and, as in genetically normal individuals, the appearance of such a tumour is nearly equivalent to a death sentence. Recurrence and generalization of such cancers is the almost invariable rule. It is a hard thing for any human being to be condemned to such a life, with face heavily pigmented and distorted by scars, often with partial blindness from corneal ulceration, a taboo on direct sunshine, and the continual appearance of those potentially lethal lumps in the skin. Yet the only patient I have seen, a woman in her thirties, had remained cheerful about her afflictions, and seemed even to get a little pleasure out of the intense interest physicians and dermatologists were taking in her clinical condition.

As an Australian living in a sunny country, I am deeply interested in skin cancer, which is one of the characteristically Australian diseases, particularly in Queensland and other parts of tropical Australia. Almost any north Queensland sugar grower or other outdoor worker can expect to find one or other type of skin cancer on face or hands during his old age. Its incidence has been clearly related to the duration and intensity of the patient's exposure to direct sunlight during his working life, and much work has been done on the types of cancer and the best means of treating them. There are four sorts that are relatively common, which we can call B, S, A, and M. Technically, they are basal cell carcinoma, squamous epithelioma, acanthoma, and malignant melanoma. In addition, there are some rare tumours—sarcoma and angioma—that can be lumped together as others (O). What is of special significance is that the frequency of the tumours in Queensland is in the order B, S, A, M, O—that is, in outdoor workers in their sixties and seventies after a lifetime of exposure to tropical sunlight. The same order, B, S, A, M, O, holds for the dozens of tumours arising in young X.P. patients, most of whom are carefully shielded from direct sunlight. The fact that the order is the same in both series indicates strongly that the same process must be at work in both, and that the ultraviolet component of sunlight in involved. With very inadequate data I have tried to work out the relative effectiveness of a given amount of ultraviolet (UV) light in producing cancer in X.P. cases, compared with normal persons, and concluded that the former must be at least

10 000 times as susceptible. This is something to be kept in mind when we are thinking about genetic disease, about sunburn, or about the nature of cancer.

X.P. cases do not all conform to the standard clinical picture that I have described. There are some fine technical differences, which do not need discussion here, but in addition we find a group comprising 30 or 40 per cent of cases with the typical skin changes and a coexistent pattern of congenital brain damage, usually evident at birth. These children have small heads, with brain atrophy and mental retardation plus a variety of other nervous symptoms. Such evidence of brain damage cannot be just coincidental; it is firmly tied to the other symptoms and signs in the skin that define a case of X.P. and must represent the activity of a genetic error carried on the same chromosome, and probably on the same gene.

The elucidation of X.P.

In this discussion of the basis of X.P., it will be impossible to avoid covering some of the same ground as in the more general account of somatic mutation in earlier chapters. Here, however, we are concerned with a specific example, and if the way the characteristic signs of the disease develop is to be understood, a little repetition here and there of points made previously should be helpful.

The first evidence that DNA could be repaired came from work done around 1948 on bacteria and bacterial viruses that had been inactivated by UV light. In suitably arranged experiments exposure to *visible* light could revive apparently dead bacteria and even in the dark there was a progressive recovery. Further study of such phenomena indicated that the UV acted by damaging DNA, and eventually it was worked out in detail how the DNA was changed by UV and how the damage could be repaired in several different ways.

For someone who is not interested in molecular detail, the essence of the situation is that a segment of a DNA strand, often quite a long one, containing the damaged units is excised and the gap refilled with newly synthesized nucleotide units. These units must be linked into the chain in their proper sequence by matching them with the nucleotide units on the complementary DNA strand. This may require some quite complicated manoeuvres in what is called replication repair, and particularly under these circumstances error is liable to creep in. By error we mean that an incorrect mucleotide or sequence of nucleotides is inserted into the DNA from which descendant cells will derive their genetic character. Any influence of the error on the functioning of the cell will be carried on to al' descendants.

With the fairly obvious role of UV in producing X.P. damage to the skin, it was natural for molecular biologists to study the capacity of cells from X.P. patients to see how their response to UV might differ from that of similar cells from normal children. As was expected, Cleaver and others found evidence of inefficiency in the process of repair. There was a delay in the insertion of new nucleotide units in place of those removed which appeared to depend on some weakness of the initial process of excising the damaged unit. As increasing numbers of X.P. patients are investigated, several distinct types of inadequacy of DNA repair processes have come to light. All seem to produce functionally similar patterns of errors in the relevant genes when repair is complete. An important point to be remembered is that mutations with proliferative effect can only arise from cells in which DNA repair is physiologically complete. Unless this is so, cell division is impossible, but an *informational* error is quite possible in a DNA strand which in the physiological sense is quite normal.

Another way of showing the difference between X.P. and normal human fibroblasts provides some interesting evidence of the basic similarity of DNA, whether from virus or man. A suspension of a DNA virus (Adenovirus) is exposed to UV and then titrated by growing it on cultures of normal human fibroblasts in parallel with cell cultures from the X.P. patients being studied. So far, all the X.P. cultures tested show lower counts of living virus units than are obtained with fibroblast cultures from normal people. This is interpreted as demonstrating that virus DNA damaged by UV light can only be repaired by the DNA-handling enzymes of the host cell that the virus infects and can multiply in. Any exceptionally error-prone enzyme will fail to complete repair of the damaged virus-DNA, with the result that X.P. cells will repair and resuscitate fewer UV-damaged virus units than normal fibroblasts.

The overall interpretation to cover both the experimental studies on DNA repair and the changes in the patients is that there is a weakness, an error-proneness, of genetic origin in some element of the complex of enzymes that repair DNA damaged by UV radiation. As in Chapter 2, we speak of this complex, in terms of its most extensively studied enzyme, as DNA polymerase, or DP. There is, however, definite evidence that several types of genetic damage within the complex can give similar types of error-proneness in the repair process. What happens is basically similar to the interpretation given for ageing in an earlier chapter, but the degree of error-proneness is very much greater. I have already suggested that it might be 10 000 times the normal level. Such a degree of error-proneness allows us to look more closely at the nature of somatic mutation.

Somatic mutations

A somatic mutation is a recognizable change in chemical structure or cellular functioning resulting from genetic error in the DNA of a somatic cell. It is axiomatic that the change is also expressed in any descendants of the mutated cell. Informational error resulting from the insertion of a wrong sequence of nucleotide units can have a variety of results on the cell, and most of these are more readily recognized in X.P., with its excessively error-prone enzyme, than in any other system. The commonest result of UV light damage to cellular DNA is probably complete error-free repair. When extensive damage is done, physiological repair may be impossible and the cell dies. Even when repair is complete, the informational errors can be lethal and again the cell dies. Both results will be much more common in X.P. cases than in normal people, and the death of large numbers of epidermal cells is responsible both for the acute inflammatory reaction to sunlight and the later atrophy of skin structures. With a viable cell containing informational errors in repaired DNA there are many ways in which the cell can be changed, but from the medical point of view most of them can be forgotten, since they produce no visible effect. Only when the error results in abnormally rapid proliferation of the cell's descendants will the existence of the error be noticed. It by no means follows that in a given type of cell only one form of proliferative response is possible. This is particularly evident in X.P. Normal pigment cells (melanocytes) in a European are only lightly filled with pigment granules and they are very uniformly distributed in a layer just beneath the epidermis. The mutant cells in a freckle or mole are more heavily pigmented, and descendants of the original mutant spread peripherally. A characteristic of X.P. is that close examination will show freckles with a wide variety of size, shape (round or irregularly star-shaped), density of colour and its uniformity over the pigmented area. Moles are accumulations of melanocytes more than one cell layer in depth. They too vary in size and appearance and may change almost inconspicuously at first to the highly lethal malignant melanomas. No doubt finer structural and biochemical study would uncover other differences between the various clones of melanocyte that make up the individual pigmented patches, either freckles or moles. However, the point already mentioned elsewhere and particularly evident in X.P. is that there are many ways by which mutation in a single class of cell can be manifested. It may give rise to an innocent proliferative change such as a freckle or to a tumour like malignant melanoma whose cells have a capacity for growth in culture and malignant attributes in the body.

Epidermal cells (non-pigmented) have similar capacities. There are innocent wart-like pilings up of superficial cells which are spoken of as hyperkeratoses, as well as mildly malignant acanthomas and basal cell carcinomas, and more malignant squamous cell carcinomas. Each breeds true to type, and within each group individual characteristics of some of the tumours can be observed. It it typical of an error-producing process that the observable cellular change which results should be individually distinct from clone to clone, reflecting the changes in the mutant cell from which the clone arose.

As I shall elaborate in the next chapter, cancer is not a strictly determined process giving rise to a standard form of proliferative and invasive cell. It is better understood as resulting from a wide variety of genetic errors in germ-line and/or somatic cells and usually requiring a sequence of two or more errors, one of which is always in the somatic cell from which the tumour is immediately derived. The errors involved may have many distinct forms, and probably every primary tumour is to some degree different from any other. The more measurable characters one takes account of, the more often any given tumour will be differentiable from others arising from the same cell type.

A variety of genetic anomalies in the central nervous system may occur in association with typical X.P. symptoms in a proportion of cases—and consistently as the so-called deSanctis-Caccicone syndrome in certain families. Since the lesions are present at birth, there can be no question of their being due to UV damage to DNA with informational error during repair. Eventually it may be possible to define the nature of the disease-producing process in the central nervous system, but at present we can only speculate that the same error-prone enzyme complex that is responsible for the skin mutations fails to act correctly in some DNA functions that are concerned with the process of neural differentiation. This is pure speculation, but if some wholly different process is responsible, then we shall need to recast the whole theoretical basis of disease inherited according to simple Mendelian formulae, such as that applicable to autosomal recessive conditions. If we accept the orthodox view that when a condition is inherited in Mendelian fashion a single gene only is involved, all features of the disease that are regularly present must be due directly or indirectly to changes in that gene. This contention has recently been justified experimentally. Cells from patients with these neurological signs as well as X.P.

were more susceptible to being killed by ultraviolet than cells from standard X.P. patients. Another variant of weakness in DNA repair has been uncovered, and in some way this must be related to the occurrence of brain damage.

The broader significance of X.P.

When the first comprehensive study of X.P. appeared in early 1974,[1] I was excited by this explicit evidence for the existence of an excessively error-prone system of DNA repair which seemed to have nearly all the qualities that I had postulated a year previously for the much less error-prone system to which somatic mutation and its derivatives, ageing and malignant disease, were ascribed. Subsequent consideration[2] has only slightly weakened the importance of X.P. as a support for the intrinsic mutagenesis hypothesis. The relevant points can be enumerated.

1.   If varying degrees of error-proneness in the repair and replication of DNA were responsible for differences in lifespan and the time at which cancer appears, all experience in medicine would lead one to expect that occasionally these differences would be exaggerated. There *might* be virtually error-free systems and there should certainly be others that were excessively error-prone. The latter should produce changes that could within reasonable limits be equated with those of ageing and cancer.

2.   Again in line with experience, the last thing to be expected is a simple acceleration of ageing and early appearance of cancer. Any body system evolved to handle the normal eventualities of life is a highly complex self-regulating mechanism calling for the action of large numbers of genes. Any Mendelian condition involving only one gene will give an unbalanced anomaly of some sort.

3.   In X.P. the gene defect can find possibilities for expression only in the tissues susceptible to UV damage, i.e., the exposed parts of the body, with, in a proportion, a still obscure action on the process of brain development before birth.

4.   The early and multiple occurrence of skin cancers of various types makes it impossible to estimate whether there is any 'normal' shortening of lifespan in X.P. cases. The only relevant point on record is that the skin on light-exposed areas shows atrophy and other evidence of changes similar to those occurring in old age.

5.   The very great increase in the rate of tumour formation with the retention of the same types and distribution of tumours as are found after very long exposures to sunlight in people with, we believe, DNA-handling enzymes of a much lower degree of error-proneness may represent the strongest evidence yet provided

on the nature of the processes most often concerned in the initiation of cancer.

As will be discussed more fully in the next chapter, my interpretation of malignancy is that both replication of normal DNA and repair of DNA damaged by physical or chemical agents is liable to give rise to informational errors in either structural or regulatory genes. The likelihood of any identifiable type of mutation—the expression of such an informational error—will depend essentially on the extent and complexity of the damage to be repaired and the error-proneness of the enzymes concerned with replication and repair. This would hold for the types of somatic mutation that can be recognized as cancer and those that by their accumulation cause the changes that form the manifestations of old age. Much more is needed for a full explanation of either set of phenomena, but as a first approximation this concept makes sense of both and is very strongly supported by the experiment of nature which is responsible for xeroderma pigmentosum.

Are there any other genetic diseases similarly caused?

The recognition that the symptoms of X.P. could be ascribed to the action of genetically determined error-prone DNA-handling enzymes stimulated a search for other human diseases that might be similarly caused. So far not very much has been achieved. By looking through the standard catalogue (by McKusick) of recorded genetic diseases of man, I could note at least twenty-one conditions that might be considered as being, like X.P., dependent essentially on the error-proneness of DNA repair and replication. Clues to be sought included such features as undue prevalence of cancer or leukaemia in early life, a multiplicity of discrete primary lesions, abnormality of the immune system, and signs of premature ageing. A range of conditions is described as having one or other of the three Mendelian types of inheritance and presenting one or more of these marker symptoms. Most of them are rare or very rare conditions and none have yet been unequivocally shown to be associated with DNA repair anomalies.

One comes on a human dilemma at this point. Scientific curiosity, once it has been developed and found to bring enlightenment and professional prestige, can be nearly as obsessive as the drive for power in politics. In medicine the experiments of nature that I have been talking about represent, for the patient, a desperately unfortunate and undeserved mischance that has left him with a life that, it he had known anything else, would be utterly intolerable. On the side of the physician engaged in clinical research there will be, depending on his temperament, a variable degree of kindness and compassion for the patient but inevitably an intense interest in the abnormalities he presents, and a determination to understand how they have arisen will dominate his attitude. In all probability, since these are genetic problems, nothing that he will learn about the condition will provide more than minimal help toward effective treatment. The justification for deep study of rare genetic conditions is to help clarify the complex means by which gene action is expressed in the tissues and somatic cells of the body, plus the reasonable hope that in one way or another a proportion of such investigations will provide a little knowledge that can be utilized for more direct human benefit.

Although I have had no direct contact with the National Institutes of Health hospitals in Bethesda, Maryland, the approach adopted there for the investigation of rare human diseases impresses me as probably the most satisfactory compromise between the needs of the patients and the quest for new knowledge in clinical science. Physicians in most of the teaching hospitals of the country are asked to refer patients with certain specified rare conditions to the N.I.H. The patient is assured of first-class hospital care and treatment with no financial cost to himself in return for his willingness to undergo the minor discomforts of much more intensive investigation than would be his lot in a normal American hospital. In this way it is possible to collect together ten or twenty patients with even extremely rare conditions and obtain a much clearer outlook on the disease than would be possible when, as is usual in most hospitals, only an occasional patient is seen at intervals of some years. I have not read a patient's account of the N.I.H. experience, but I should guess that in most ways it is the best solution available for a basically distressing situation. It is human to feel a certain sense of gratification to be a centre of interest for a group of people and to have the sense that if anything can be done to alleviate one's disability it will be done here. From the side of the physician interested in how medical science is developing, I can testify to the enlightenment I have gained about X.P. and another rare condition that interests me—Sezary's disease—from reviews of such N.I.H. work.

I have no doubt that in the next decade or two, similar studies of some of the rare conditions among the twenty or thirty candidate diseases in McKusick's catalogue will come from the N.I.H. or from equivalent centres elsewhere.

Most of the conditions that I included in my list are so rare that many physicians will have forgotten that they exist and even well-informed laymen will probably have heard of only one or two.

However, as a gesture of good faith and of possible help to some interested reader, the following is a list of conditions in which one may suspect some genetically based error-proneness in handling DNA rather than a classical defect in a structural gene. The twenty-one conditions mentioned are divided into three groups, in only the first of which is the evidence for such an origin relatively strong. In this and the two other groups the serial number in McKusick's catalogue is shown against each disease.

Group I: Generalized conditions

McPherson type of xeroderma pigmentosum (a mild form inherited as a dominant condition)	19440	Progeria	26410
		Thrombocytopenia, absent radius syndrome	27400
Ataxia telangiectasia	20890	Werner's syndrome	27770
Bloom's syndrome	21090	Wiskott-Aldrich syndrome	30100
Cockayne's syndrome	21640	Congenital dyskeratosis	30500
Fanconi's pancytopenia	22790		

Group II: Conditions with multiple skin lesions

Basal cell nevus syndrome	10940	Heriditary multiple leiomyomata of the skin	15080
Hereditary multiple benign cystic epithelioma	13270	Multiple neuro-fibromatosis	16220
Self-healing squamous epithelioma	13280	Porokeratosis of Mibelli	17580
		Tuberous sclerosis (Epiloia)	19110

Group III: Conditions associated with accelerated fallout of neurons in parts of the central nervous system

Alzheimer's disease (Pick's disease)	10430	Amyotrophic lateral sclerosis/ Parkinsonism-dementia of the Chamorro people of Guam	10550
Amyotrophic lateral sclerosis	10540		
		Huntington's chorea	14310

10

Age-Associated Disease

When a man or woman dies in old age the cause or causes of death will fall into one or more of three categories. The first (A) includes the relatively trivial accidents or infections that can cause death in a person made highly vulnerable by old age. The second (B) covers those causes that would be fatal in any adult and have no special relationship with age. The third group (C), with which we are concerned in this chapter, are those that are commonly regarded as being clearly associated with old age: cancer and cerebral haemorrhage, for example. Most deaths in the old can be ascribed to causes grouped under A or C, and in almost every instance careful study would show some contribution to the fatal outcome from *both* A and C.

Definition of age-associated disease

Before trying to define what constitutes an age-associated disease it is best to start with a closer look at group A, those conditons which, although regularly associated with old age, are clearly manifestations of the greater vulnerability of the old. Perhaps the best indication is to look at the causes of death that are more frequent in infancy and early childhood and in old age than in persons between human five and fifty-five years of age, i.e., conditions to which both extremes of life are vulnerable. Pedestrians killed in road accidents and persons dying from respiratory infections during an influenza epidemic show such a distribution of ages. In general, the relatively non-specific infections of respiratory tract and intestine, and accidents that would be regarded as trivial in a younger person, are the important components in group A.

With those out of the way, the best initial approach to the age-associated diseases is to look at the way the number of deaths from a certain disease is related to age. Such information is readily available in the demographic statistics of all advanced countries. If we are interested, say, in cancer of the breast, we shall find for each year the number of deaths in women aged 21 to 30, 31 to 40, and in ten-yearly groups until the list terminates often with 71 to 80, and over 80. To make use of the data, we need in addition to know what proportion of the total population is to be found in each of those age brackets. The age-specific incidence of the disease at, say, 51 to 60 is obtained by taking the number of deaths in women reported as due to breast cancer during a given year, dividing this by the number of women in this ten-year age bracket who were alive at the beginning of the year, and multiplying the result by 100 000. As cancer of the male breast, though very rare compared with the female condition, is by no means unknown, a similar estimate will be made in terms of the male population. Having obtained figures for each ten-year age

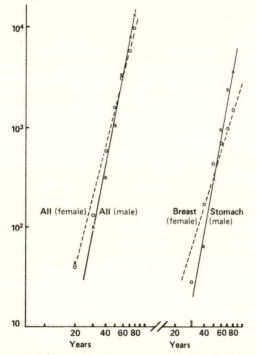

Fig. 6: *Representative specific age incidence curves for human cancer: all cancers (male and female); cancer of stomach and duodenum (male); cancer of the breast (female). Figures from Australian experience (Lancaster).*

group and for both males and females, the results are plotted on graph paper: age incidence per 100 000 vertically (ordinates); ages horizontally (abscissae).

It is a human weakness to think that straight lines on a graph are nice to look at and easier to understand. So it is universal to plot the incidence logarithmically so that each major division represents a tenfold increase over the one beneath it, as shown in Figure 6. If we are dealing with all deaths or with deaths from infections of any sort, the ages are plotted linearly, each ten-year interval being the same. This results in an approximate straight line, starting at 21–30 and rising steadily. With cancers and other non-infectious conditions it is more usual to make the age scale also logarithmic so that along the horizontal axis we have equidistant marks for 10, 20, 40, and 80 years with the intervals subdivided logarithmically. For most cancers and for a number of other conditions this gives something close to a straight line.

An age-associated disease, then, can be defined as one whose specific age incidence curve rises steadily and usually quite steeply from its beginning some time in the twenties or later, to the 61–70 age bracket at least. After 70, numbers of people still alive fall rapidly, and some special considerations concerned with the likelihood of small genetically resistant subgroups arise. The two important groups of fatal age-associated diseases are, first, malignant disease, including cancer, sarcoma, and leukaemia, and, second, the various forms of cardiovascular disease that depend on changes with age in heart and arteries.

Cancer

Ever since Koch proved in 1876 that the anthrax bacillus was the cause of that disease in cattle and sheep, and Pasteur five years later demonstrated to all the world that a vaccine would protect against it, there has been a persisting dream that in some more or less equivalent fashion the cause and a means of preventing or curing human cancer will be discovered. The current version of the dream depends on the fact that a number of viruses that have been obtained from various sources by laboratory manipulations, when injected into an appropriate host—often a special strain of mice but sometimes guineapigs, rats, or chickens—produce leukaemia or some form of solid cancer. The facts are unequivocal and the hopeful implication is that eventually one or more viruses responsible for human cancer will be isolated and appropriate vaccines prepared to prevent and treat the common forms of human cancer. There are a number of laboratory workers who would say that substantial progress has been made in this direction, and probably most physicians who specialize in cancer look forward to their being right.

I have not personally worked with cancer patients since I was a hospital intern fifty years ago, nor have I done any personal work on laboratory aspects of cancer. Nevertheless, I have written extensively on the subject and have followed with some assiduity what has been written by others. Since 1957 at least, I have looked on malignant disease as resulting in all cases from genetic error (somatic mutation) in one or a very small number of body cells. Mutation can occur spontaneously or as a result of damage to DNA by physical or chemical means followed by repair. This holds as much for cancer as for the other mutations discussed already in Chapter 2. Relatively mild non-destructive infection of a cell by a virus can in principal and experimentally induce DNA damage and mutation (or trans-formation) of the cell to malignancy. Nevertheless there is no evidence that any of the common human cancers are produced by viruses, and in a paper published in 1957[1] I expressed deep scepti-cism about the significance for human cancer of the rather numerous 'cancer viruses' that were being studied then in mice and chickens. I felt that work along such lines was likely to be much less fruitful than investigations based on the hypothesis of genetic error and, on the whole, subsequent experience has supported that view. In my opinion there have been only two outstanding discoveries about the nature of cancer since 1957. They are the association of foetal or embryonic antigens with some types of cancer, and the enormously accelerated production of skin cancers in xeroderma pigmentosum. Both are compatible with the 1957 formulation. The fact that the approach has stood up to twenty years of intensive research on cancer cannot prove that it is correct, but it at least justifies my discussing human cancer with no more than an occasional sidewise glance at viral possibilities. This is primarily a book about ageing, and one of the most important objections to a virus cause of cancer is that all the common cancers have an age incidence utterly unlike that of any virus disease. This is the more striking since the only two human cancer-type diseases, both atypical, that are definitely associated with viruses—common warts and the Burkitt lymphoma of central Africa—have age incidences similar to other types of virus infections.

Statistics of human caner, national and international, are avail-able in profusion in every medical library and public health department in the civilized world. Evidence on the incidence of cancers in other species of mammal is much more difficult to find. Plenty of statistics exist about the tumours that are found in the course of inspection of animals slaughtered for food, but most of these are very young animals and none are more than equivalent to middle age in man. We therefore know a good deal about the rare tumours of young domestic animals, which correspond to the equally rare group of tumours of infants and young children. That, however, is a small, special group, very important but unrelated to the problem of age-associated cancer.

Evidence about the natural incidence of cancer in animals is also available from the experience of veterinary hospital surgeons dealing with horse, dog, and cat ailments, as well as tumours in cattle, especially the well-known eye cancer of Herefords. Data from a review by Priester and Mantel[2] shows a sharp increase in incidence in older animals of all four species. Figure 7 depicts the results of an experiment[3] in which 400 wild mice were caught and maintained until death under good laboratory care. The incidence of malignant

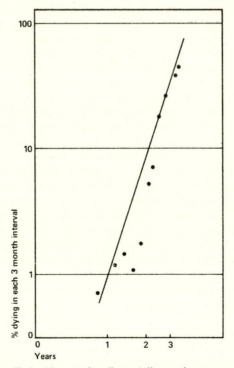

Fig. 7:   *Age-specific incidence of malignant disease in a group of 400 wild mice kept under laboratory conditions. Log-log plot of smoothed data for comparison with Fig. 6.*

tumours, mostly lymphomas, also followed a similar age incidence in these mice. None of the information from animals is as reliable as could be wished, but it is enough to show that other mammals resemble man in suffering from cancer of most types as an age-associated disease.

The most significant feature of this comparison of cancer among mammals is that, despite the wide range in average lifespan from species to species, malignant disease becomes conspicuous at about the same time as the average lifespan is approached. The nearly straight-line graph with a roughly comparable slope is seen in them all. One can hardly escape the conclusion that some common mechanism is at work in all six species and that whatever produces cancer seems to have some important association—maybe even identity—with the cause of ageing in mammals.

The natural history of cancer

It used to be said that during the last quarter of the nineteenth century more was being written about tuberculosis and the tubercle bacillus than about any other human disease. Since the disappearance of tuberculosis as a killing disease in affluent countries, its place has been taken by cancer. There is a steady flood of information pouring into the libraries, and the only difficulty is to choose what is relevant to one's own special topic.

The theme in this chapter is essentially a search for ways by which the age association of malignant disease can be brought into a logical association with the general theory of ageing. This necessarily means a concentration on some aspects of cancer research and a relative neglect of others. In the generally agreed absence of evidence of significant virus findings in any human cancer except Burkitt's lymphoma, I shall omit completely any account of work with cancer viruses in animals. I know that I am to a certain extent dodging the issue by dismissing cancer viruses as laboratory artifacts of no relevance to the understanding of human cancer, yet no elaboration and qualification of that statement would provide anything additional that was relevant to our particular theme.

In discussing the natural history of mammalian cancers, it is convenient to differentiate them into three groups. They are, first, the common epithelial tumours of skin, stomach, breast, uterus, bowel, and lung, technically carcinomas and the usual type of disease that the word 'cancer' calls to mind.

The second important group are those derived from cells that circulate in the blood during some period of their development. When such cells become malignant, they will usually predominate in the blood and be recognized as a leukaemia, but in all leukaemias some solid accumulations of tumour cells are present, and one may also see solid tumours with minimal involvement of the blood. Malignant cells may arise from lymphocytes, the blood cells that are mainly responsible for immunity, or from the so-called granulocytes, which include the actively phagocytic cells that accumulate as pus cells in acute local infections. With some exceptions, malignant change either takes place in relatively undifferentiated cells or results in a change toward the undifferentiated form.

Recent research in immunology (see Chapter 6) has allowed us to divide the lymphocytes in the body into two groups according to whether they are differentiated in the thymus (T cells) or in the bone marrow (B cells). Pathologists and physicians are now becoming interested in the role each type plays in disease and a good deal is already known. The commonest conditions are chronic lymphocytic leukaemia, lymphomas and reticulum cell sarcomas from B cells, and multiple myeloma from the mature plasma cell form of the B cell. T cells can also give rise to rarer tumours and leukaemias with the unfamiliar names of mycosis fungoides, and Sezary's syndrome. In addition, there is a group of three conditions in which lymphocytic proliferation with some malignant features is associated with known or presumed virus infection: Burkitt's lymphoma, Hodgkin's disease, and infectious mononucleosis.

These epithelial cancers (1) and what can be called leukaemias and lymphomas (2) are the important forms of malignant disease and are the only ones that will be elaborated. The remaining forms, which will merely be mentioned, include sarcomas from cells of connective tissues, teratomas from reproductive cells, and a group of tumours arising from immature cells in early life; the eye tumour retinoblastoma is the prototype of the group.

Monoclonality and related features of cancer

A cancer or a leukaemia is a mass of cells, all appearing very similar under the microscope, and in almost every significant way in which cell quality can be tested all the cells of a single cancer seem to be uniform. As I discussed in regard to monoclonal conditions in Chapter 5, some of these tests depend on the individuality of the normal cells in the tissue from which the cancer arose, and with such tests it is possible to say that certain tumours must have arisen from a single cell. They represent the clone of descendants from one cell that underwent an inheritable change, and we speak of the population of abnormal cells as being *monoclonal* in character. Where conditions are not complicated by extraneous factors, it is nearly always found that well-defined malignant tumours and leukaemias are monoclonal.

This has the important implication that what starts a cancer on its uncontrolled proliferation involves one cell only, a cell that is different in that respect from all its millions of associated cells in the tissue concerned. Now, apart from human interference with a micromanipulator, the only thing than can happen to one cell in a hundred million that *could* be recognized in all its millions of subsequent descendants is an informational error in the DNA of that cell—a somatic mutation. Simply on the evidence of monoclonality one must hold that, whatever other 'causes' of cancer there may be, such as cigarette smoking, exposure to ultraviolet light, or infection by a cancer virus, the immediate cause is the informational error in one initiating cell. The point has already been raised in relation to xeroderma pigmentosum (X.P.), where the cancers are of precisely the same type as those that occur in elderly normal people. What is called for here is some further elaboration of the process by which cancer is produced by carcinogens like the tars in cigarette smoke and the thousands of organic compounds which on injection into animals will produce cancer.

It has now been shown convincingly that all those carcinogens (cancer-producers) that are suitable for testing can act as mutagens (mutation-producers) on certain bacterial strains susceptible to mutate in fashions that can be conveniently measured. Former doubts about such identity depended on the fact that when a certain chemical appeared to be an active carcinogen but was not mutagenic, the real carcinogen was an activated form of the original chemical, produced by the action of enzymes in the mouse or other test animal's tissues. In very similar fashion to what occurs in ultraviolet damage to DNA, the activated carcinogen reacts with DNA in the nucleus, causing regions of damage which must be recognized, excised, and repaired by processes of the same type as are employed in replication and in other types of repair. Again, depending on the extent and difficulty of the repair and the error-proneness of the DNA-handling enzymes, informational error will be liable to occur. If other conditions are right—and they may be highly complex—the informational error will be perpetuated as the malignancy of a clone of cancer cells.

Studies on experimental animals, using much larger doses of carcinogenic chemicals than would be likely to be concerned in human situations other than cigarette smoking, have shown that the rate of cancer production is related to the time over which the carcinogen is applied and to its effective concentration or dose.[4] This signifies that errors arising during repair of the damage done by the carcinogen-mutagen are much more important than the errors that might arise during normal DNA replication. Evidence summarized by Sir Richard Doll points to a similar conclusion for the relation between cigarette smoking and lung cancer in man. The likelihood of clinical lung cancer is proportional to the number of cigarettes smoked per day multiplied by the number of years the subject had been a smoker.

Genetic aspects of cancer

Although most aspects of the initiation of epithelial cancers can be covered by the postulation of genetic error following replication, or in the course of DNA damage by physical or chemical mutagens, other factors need consideration. Some human cancers are clearly inherited, notably the eye tumour of children, retinoblastoma. This, as its name indicates, involves embryonic cells from which the retina subsequently develops. The tumour must therefore be initiated before the formation of the retina is complete. Knudson,[5] from an epidemiological and genetic study of the available information, concluded that a retinoblastoma arises as a result of two consecutive mutations in a single cell. The second of these is always a somatic mutation (S2), whereas the first may be an inherited anomaly (G) representing a germ-line mutation in a preceding generation, or also a somatic mutation (S1). Where the sequence is G S2, any S mutation that occurs will result in the appearance of a tumour, so that multiple tumours are common. On the other hand, if only a few S1 mutations occur in millions of infants with initially normal retinoblast cells, the chance of a second mutation affecting S1 mutant cells will be extremely low. So the rules hold that all S1 S2 tumours are single and all bilateral or multiple tumours are G S2 in character. A proportion of single tumours will also have the G S2 character. Essentially similar behaviour is also seen with two other rare tumours of embryonic cells, pheochromocytoma and neuroblastoma.

Knudson's interpretation of retinoblastoma is almost certainly correct, but, as is the way with most interpretations of cancer, leaves a great deal still to be explained. Other types of epidemiological and statistical study indicate that there is a genetic component in the origin of many types of cancer. It is obvious from all the statistics of lung cancer that the majority of heavy smokers of cigarettes die of some disease other than lung cancer. Part of this apparent insusceptibility is of statistical origin; if all heavy cigarette smokers could avoid every other cause of death and continue their addiction at the same level indefinitely until they died of lung cancer, there would probably be very few surviving to 100. There is, however, some familial evidence of inheritance of susceptibility or resistance and to some extent this holds, as might be expected, for all types of cancer. Nearly all discussions of such problems lead to the assumption that two or more consecutive mutations are needed to allow the final change which sets a cell proliferating as a clone of cancer cells.

No one seems to have even a suggestion as to what form the preliminary mutations take or why there is such relative uniformity in the end product of two rare and random occurrences. In a rather hazy fashion one can picture a continuing scatter of errors involving minor changes in regulatory DNA in themselves of no special significance but setting the stage for a new error to trigger a major move toward active proliferation and cancer. One speculation that feels promising to me is to think of the final initiation as corresponding to the derepression (= activation) of a developmental programme requiring the cell in question to undergo a series of replications until the programme is turned off by some appropriate feedback when the proliferative phase has achieved its normal biological objective. If that feedback can never be set in action in the mature organism, the proliferation might well continue indefinitely. This is a mere speculation, but it gains some credibility from the various findings of foetal antigens in malignant tissues.

Foetal antigens

In Chapter 5, the carcino-embryonic antigen (CEA) originally obtained from cancers of the large bowel was briefly mentioned. The substance, which has now been purified and extensively studied, was defined by the type of antibody it could produce and is therefore spoken of as an antigen. Although there seemed to be no such antigen in normal tissues, it was found to be present in extracts of liver and other abdominal organs from human foetuses. With the development of increasingly sensitive methods of detecting the antigen, the situation has become progressively more complicated. The antigen is present in the blood in most cases of active cancer of the bowel and soon disappears after surgical removal of the tumour. Once this had been discovered, the obvious thought was that here was a test for the early diagnosis of bowel cancer that could allow early and effective surgical removal of the tumour before it had spread to other organs. Cancer of the large bowel rather often gives no symptoms at all until it has become extensive enough to block any movement of contents along the bowel. Then the acute symptoms of intestinal obstruction suddenly make surgical exploration urgent and the tumour is found more often than not with secondary nodules in the liver, and no hope of cure. A test that would detect a tumour when it was just a small lump in the bowel wall could be a lifesaving advance.

Unfortunately there are difficulties in all directions. Middle-aged and elderly men and women who consider they are quite healthy, with regular bowel function and no abdominal pain, are not interested in having a few ccs of their blood taken for a cancer test. In any case, to test everyone in the 55 to 75 age group once or twice a year would be quite enormously expensive; a whole army of technicians would have to be trained before it could even be contemplated. Then there are weaknesses in the test for CEA itself. Almost certainly a single small tumour would give a negative or equivocal result, while rather weaker but still positive results can be obtained from a variety of other tumours and some non-malignant conditions, such as cirrhosis of the liver. It has even been reported that heavy cigarette smokers with no evidence of cancer may be positive.

Another well-defined antigen, α-foeto-protein, present normally in foetal blood, appears in most types of liver tumours in adults and may be detected in the patient's blood. In recent years, too, many reports have been published about other types of cancer antigen that are common to some type of foetal or embryonic cells of the same species. Not all these findings are fully substantiated, but they all help to support the hypothesis that cancer may often represent the turning on, by some combination of genetic errors, of a programme appropriate to some type of prenatal development. Lest this should seem too farfetched for some readers, it may be advisable to repeat the axiom that within every somatic (or reproductive) cell nucleus all the information coded in DNA that is necessary for the construction and maintenance of the body, from the first activities of the fertilized ovum to old age, is potentially present. Each segment of information is always there, waiting to be actualized when required or when some rare and accidental error intervenes.

One other characteristic of cancer arising from epithelial organs is the capacity to develop a whole series of changes after the tumour has become clinically demonstrable. A cancer is an unco-ordinated mass of multiplying cells often rather poorly supplied with blood and no doubt lacking many of the refined metabolic processes of any normal tissue. Under such circumstances there must be innumerable possibilities for error to arise in the repair of damaged DNA or its poorly co-ordinated replication. Many mutants would be expected to appear and to become involved in an unending struggle for survival. It is as Darwinian a situation as the swarming mixture of micro-organisms, maggots, and other creatures that putrefy and eventually dispose of an animal carcass. At every stage some successful mutant gives rise to a dominant clone which gives its character to some advancing pseudopod of the tumour or some secondary metastasis to another part of the body. This changing character of a cancer, as it spreads or is transplanted experimentally from one new host individual to another, is spoken of as 'progression'. It can be demonstrated in many ways.

In nearly every cancer hospital anywhere in the world there is no lack of patients with cancer of the lung—cigarette cancer—and there has been much clinical interest in a variety of symptoms that seemed to have nothing in common with the expected symptoms of lung cancer. The most striking group of symptoms suggested that one of the hormone-producing glands, usually the parathyroids or the pituitary, was liberating excessive amounts of a hormone into the blood. In other patients there were symptoms of autoimmune disease that might affect the brain, the joints, or the skin. I remember well the story of a woman admitted to a New York hospital with exceptionally active and severe rheumatoid arthritis. Routine X-rays showed an unmistakable lung cancer, which was removed surgically. Within a week or two the symptoms and positive blood tests of rheumatoid arthritis had disappeared. But they came back in force a year later, with the eventually fatal recurrence of the cancer.

The accepted and undoubtedly correct interpretation is that this is an ectopic—meaning in the wrong place—production by the cancer cells of a protein that they should never have synthesized. Ectopic protein production is the strongest evidence from human material that somatic cells, as I have already said, contain the whole library of information that is present in the fertilized ovum, but under normal circumstances most of the volumes are firmly sealed; only those specifically needed for the particular cell are accessible for use. In every cancer there has been a breakdown in law and order between and often within its cells; every type of error becomes possible, and even firmly locked volumes are broken open and put to random use. There are many 'wrong' proteins that may be produced by cancer cells unknown to anyone because they produce no signs of their presence. Only when the aberrant protein liberated is a very potent hormone with an unmistakable action or a protein that can act as an autoantigen and provoke recognizable symptoms are we able to say that ectopic production has taken place.

Chromosomal changes

An even more illuminating set of phenomena that points in the same direction emerges when we study changes in the number and form of chromosomes as a cancer develops and spreads, or is transferred experimentally in the laboratory. The chromosomes of a cell nucleus take on their characteristic form at only one stage in the cycle of cell replication, but there are now well-established technical methods for stopping the process at the right point and for counting and classifying the individual chromosomes. In normal human beings, each cell has forty-six chromosomes—twenty-two pairs of autosomes, each with its own size and staining qualities, and the two sex chromosomes, XX in females and XY in males. In describing what happens to the chromosomes in cancer, I am going to combine what has been found in some human malignancies with much more intensively studied material from animal cancers.

When a cancer first appears, its chromosomes are normal in number and appearance, but by the time it is clinically recognized and examined, one usually finds a certain degree of abnormality: an extra chromosome or an abnormally shaped one where part of one chromosome has become attached to another can be taken as examples. These are spoken of as marker chromosomes and can be used to identify different clones of cells. At this relatively early stage it is common to find one marker dominating the picture and identifying the clone that has been able to proliferate most successfully. When a mouse or rat tumour is transferred experimentally from one susceptible animal to another, a constant sequence of proliferation, mutation, and selection is going on within its cell populations. Detailed study of the chromosomes at appropriate stages allows us to see the appearance of new clones which may become dominant for a while before being replaced in their turn by some new mutant. With each change the chromosome pattern deviates further from the normal and sometimes may become completely unrecognizable as a human (or mouse) pattern, often with quite abnormally large numbers of chromosomes or fragments of chromosomes. In these late stages, many associated errors of all sorts in DNA must be present and in all probability highly error-prone enzymes are constantly appearing, giving rise to subclones that rapidly become non-viable. In the limit, only those qualities that are essential for continued proliferation will be left.

Cancer is truly a black apotheosis of biological error.

Leukaemias and lymphomas

The second group of cancers that seem to be of general interest for a dissertation on ageing in man are those that derive from cells that normally circulate in the blood. The commonest arise from the lymphocyte in one of its several forms, but there is an important leukaemia related to the cells from which the common phagocytes of the blood arise and some rare conditions in which red cells, eosinophils, monocytes take on malignant or near-malignant qualities. The lymphocyte is the basis of immunity, and I have mentioned something about lymphocyte tumours and leukaemias in that contest in Chapter 6. In this chapter I want to concentrate on tumour aspects of the leukaemias and solid lymphomas derived from T and B lymphocytes. Immunologists are still far from clear about the detailed origin of T and B cells and their subclasses, but the general teaching is that a common stem cell is directed toward T or B, according to where or how it is stimulated to differentiate. B cells are finally converted into specialized antibody pro- ducers—plasma cells—and there are hints of further specialization in T cells.

Taking B cells first, their malignant derivatives include: chronic lymphocytic leukaemia, a common and usually slow-moving disease of typical age-associated character; multiple myeloma, in which proliferating plasma cells produce tumours, usually in the bone marrow, and liberate large amounts of a single antibody-globulin into the blood; Burkitt's lymphoma, now rather well known as a childhood tumour, often of the jaws, which is virtually restricted to malarial regions of central Africa and New Guinea; a variety of tumours, usually spoken of as lymphoreticular tumours, or reticu- lum cell sarcomas, mostly in elderly people.

T cell conditions include a type of lymphocytic leukaemia, spoken of as Sezary's disease, in which the characteristic nuclei have a double or nearly double complement of chromosomes and an ex- panded nuclear surface which is folded to make the nucleus look like a tiny brain. This is probably only one form of a set of related lymphomas and leukaemias that have a special tendency to produce damage and tumours in the skin. Acute childhood leukaemias are probably of stem cell origin, but 10 or 20 per cent have T cell character.

Mainly because T cells are by no means as well understood as B cells, only tumours of the latter class call for comment.

Multiple myeloma is mainly known for the production of large amounts of immunoglobulin which in any individual patient is, with rare exceptions, of a single specificity, every molecule being made precisely to the same pattern, a quality which does not as a rule change through the whole period that the patient survives. No pro- gression, as described for typical cancer, seems to occur. The cells behave in such relatively normal fashion that one wonders whether myeloma represents just one specific type of mutation plus an anti-body-type reactivity with some antigen that is constantly present in the body. That, however, is an unsupported speculation.

Burkitt's lymphoma is the only malignant tumour of humans that can regularly be shown to contain either virus or evidence of past infection by the virus. It is always the so-called Epstein-Barr (EB) virus, the cause of a common disease, infectious mononucleosis, or glandular fever. Infectious mononucleosis is itself something of a problem; in many ways it resembles a leukaemia, but a leukaemia that provokes a strong T cell response, which always eliminates it. It also seems to be well established that human B cells will not grow outside the body in culture unless they are of a line of cells that at some time was infected with EB virus. Burkitt lymphomas resemble other tumours in being monoclonal, which makes it unlikely that the EB virus has more than a minor preliminary role in the disease. Current opinion seems to favour the idea that chronic malaria in children weakens immunity and allows an occasional cell to escape the normally effective control of the changed B cells.

The appearance of lymphoma-type tumours in patients whose immune system is deficient as a result either of congenital disease or from drug treatment has already been described (page 81). Taking the whole picture of these lymphocyte malignancies into consider-ation, it makes sense to suggest, as I have done, that if lymphocytes are to carry out their many and complex functions effectively they need to be able to multiply rapidly and to carry out complex manipulations of their DNA to give them the flexibility they need to control the working of the immune system and to deal with all infections as they arise. Error must be unduly frequent compared to most other types of cell, and therefore one would expect a corre-sponding increase in the frequency of malignant change.

If lymphocytes have an abnormally high likelihood of genetic error and malignancy, it is a simple necessity for survival that the immune system should have evolved some special facility for recognizing and eliminating malignant lymphocytes. On such an hypothesis any serious and continuing immune deficiency, however it is induced, will be liable to allow the appearance of lymphocytic tumours in undue frequency.

Much more could be said about these lymphocytic tumours with their intimate relationship to the immune system and their suscep-tibility to quiet (noncytopathic) infection by EB virus and perhaps the still undiscovered virus which many people think must be responsible for Hodgkin's disease. To do so, however, would require too technical a discussion to be appropriate to the purpose of this book.

Cardiovascular disease

More people die of disease affecting the heart and arteries than from any other group of diseases. These causes of death include the acute heart attacks of coronary disease and various types of chronic heart failure; strokes represent haemorrhage from or blocking of arteries in the brain whose wall has been damaged by atherosclerosis, and aneurysms are dilatations of a large artery which subsequently rupture. Both the latter are usually related to an abnormally high blood pressure. I am no cardiologist and what I have to say about these conditions will be concerned only with their possible causes and their relation to ageing and hardly at all with medical aspects.

Perhaps the most satisfactory starting point is to take the widely accepted view that there are only two important causes of lethal cardiovascular disease, atherosclerosis and high blood pressure, and concentrate on them. I am very much aware of the controversies and complexities that surround both, and I am certain that the simplified approach I am going to adopt will need infinite elaboration and modification to cover all the facts. However, I hope that what I have to say, if it is looked at biologically rather than medically, will be fairly close to the truth.

Atherosclerosis is a degenerative condition of the large and medium arteries that starts as small plaques of proliferating cells in the innermost layer of the arterial wall. As these areas enlarge, they become loaded with cholesterol and other fatty material, particularly in persons who apparently for genetic reasons have an abnormally high level of cholesterol in the blood. The affected areas may break down and ulcerate, and over the years a variety of local pathological changes, often associated with the deposition of small clots of fibrin, are liable to develop in the regions involved.

Internal damage in coronary arteries appears to be responsible for most acute heart deaths, but it is impossible to judge how far death should be ascribed to genetic factors or to the concentration of cholesterol and other fatty substances in the blood, with the subsidiary questions as to how far cholesterol blood concentration depends on genetic factors or on the fat content, e.g. proportion of saturated and polyunsaturated fats, of the diet. Out of the very large number of investigations on coronary heart disease we find a consistent excess of male deaths over female, but the difference varies from less than twofold in Japan and Yugoslavia to more than fourfold in the Netherlands and Finland. A particularly puzzling finding is that in the United States of America, the United Kingdom, and Australia, cigarette smoking appears to be an important, perhaps dominant, factor in death from coronary disease; it is not shown in figures from Greece, Yugoslavia, Italy, Finland, or the Netherlands. This is enough to indicate the complexities of the situation and the difficulty of defining environmental components in its causation.

Hypertensive disease seems to be equally difficult to sheet home to any specific cause. With the spread of the sphygmomanometer into every part of the world, an immense amount of data has been obtained and published and correlations sought with almost every environmental factor that could conceivably be regarded as relevant. There are undoubtedly islands in the Pacific where indigenes do not show the normal increase in blood pressure with age, but this is not regularly found in primitive societies and the most likely explanation seems to be that such groups are genetic isolates. Social, cultural, economic, and occupational factors have all been looked at without any outstanding influence being established. Diet has been studied intensively for many years, and sodium intake has been shown to be highly relevant. There seems to be no question that severe limitation of sodium in the diet is an effective means of lowering blood pressure. However, no natural populations with a salt intake as low as is necessary to produce a therapeutic lowering of blood pressure in Western patients have been described. Overweight is associated with high blood pressure and weight reduction is usually effective in lowering the level. Physical exercise may be beneficial only as a help in weight reduction. Excessive use of alcohol or tobacco has not been clearly incriminated.

As an outsider trying to visualize the situation by reading a few recent reviews and discounting individual enthusiasms, I feel that everything points to a basic genetic setting of blood pressure and its changes with age plus a wide range of influence by dietary and other environmental factors each of which will affect blood pressure levels according to a host of other genetic characteristics individual to each person. If it were not for the fact that there is an established relationship between high blood pressure and excess mortality, especially from strokes (cerebral vascular accidents), one might wonder whether the clinical use of the sphygmomanometer and the administration of synthetic antihypertensive drugs should not be limited to persons with definite symptoms or pathological signs directly ascribable to high blood pressure.

All the common types of cardiovascular deaths are strictly age-associated, which conforms to what we know of their causation. Both atherosclerotic and hypertensive disease can be regarded as due to progressive breakdown in the controls that should keep the hydrodynamic and metabolic functioning of the cardiovascular system at optimum level. In the last analysis all homeostatic controls are operated through sensor and effector mechanisms equipped with chemical patterns that are genetically proper to their function. Multiple genetic error accumulating with age in somatic cells provides as satisfactory an explanation for the degeneration of circulatory function with age as it does for the loss of the effectiveness of the immune system, or of the hgher functions of the brain during senescence.

11

Genetic Aspects of Behaviour

Medically significant genetic disease can be of overwhelming importance to the affected individual and his parents, but from the community angle it makes up only a small fraction of the various misfortunes that have become the responsibilities of the welfare state. The cost to the state of genetic disease in the strict sense is significant, but it is only a minor part of the whole burden of medical care. Abnormalities of behaviour are liable to occur with most of the well-defined genetic diseases of the brain, but in most parts of the world these, too, are very rare. Of much greater practical importance are the very large numbers of people of subnormal intelligence and the even larger number who for one reason or another come into conflict with society through delinquency and crime or by antisocial behaviour that does not reach the courts. Here we have the main source of human unhappiness, something that covers a whole spectrum of conditions, which range from clinical genetic diseases associated with mental retardation and/or behavioural abnormality, to intelligence defects unassociated with clinical illness, and to behavioural disorders in otherwise healthy people, which may reach the level of criminality. The problems that are presented may well be insoluble. In the first place it is always impossible to decide in any given person to what extent his condition is genetic and how far definable environmental factors have been responsible for making him abnormal or antisocial. Everyone who is sensitive to public opinion will realize that there is a common attitude that antisocial behaviour is largely an unconscious protest against an intolerable structure of society and for the rest results from a variety of environmental effects concentrated on the disadvantaged groups of the community. On the other hand there is an inescapable surge of interest in the possibility that much of the pathology of human behaviour has a strong genetic component. Environment and inheritance must both be concerned, but the balance between the two has yet to be determined. As is so often the case in human affairs, no decisive criteria of genetic or environmental causation can be given and there are the inevitable public and academic controversies on the matter. All one can attempt is to offer a commonsense basis for thinking about the degree to which genetic factors are concerned in mental retardation, psychosis, and criminality.

The immune system and the nervous system

In earlier chapters I have introduced a number of the simpler concepts of modern genetics and, I hope, have begun to build up a sense of the all-pervading quality of the genetic control of the human body and of the way in which even minute genetic error can give rise to disease. Every year the journals that review the progress of human genetics tell of new complexities in interpreting normal function and of newly recognized Mendelian or polygenic genetic diseases. A little more should be said about those complexities before we begin to think about the significance of genetics for the central nervous system and the mind that is associated with its activity.

When we are concerned with the standard approach to human genetics it is conventional to forget about two of the most important parts of the body, the nervous system and the immune system. In some ways the essence of both is their flexibility of response to a very wide range of situations, many of which will be unprecedented. They are control systems, while the rest of the body can be regarded, for our purposes, as concerned simply with the structure of the bodily machinery and its functioning under the controlling systems. In that routinely functioning part of the body there are large numbers of distinct proteins most of which are enzymes responsible for the various biochemical changes in molecules and the flow of energy that keep the body functioning. At a guess, there may be 10 000 different proteins involved, each of which is controlled down to every one of its 10 000 to 100 000 atoms, by its own structural gene, sometimes called a cistron, which is about three times the size of the molecule it controls. Errors are rare, but they occur. Most of them involve replacement of one amino acid unit in a protein by another and the majority produce no effect that can be recognized by anything but highly sophisticated biochemical tests. Only a small proportion produce serious changes. One of them, already mentioned in another context, is a change of a single amino acid in haemoglobin which gives the fatal disease, sickle-cell anaemia. A characteristic feature of modern biochemistry dependent on the past occurrence of genetic error is the very frequent occurrence of two or more distinguishable types of a single functional enzyme. Often all the types are present in the same individual, at other times they are distributed amongst the population. This multiplicity may arise by at least two genetic mechanisms, but in general all are derivatives of a single ancestral gene. One example, already mentioned, is the pair of G6PD enzymes that in African women sometimes allow us to tell whether a tumour or a leukaemia is derived from a single mutant cell. In most orthodox fields of biochemical genetics, investigators are almost wholly concerned with 'structural genes', which have such controlling responsibilities for the enzymes and other proteins of the 'routine' organs. Eventually, means will need to be developed to interpret the functional activities and limitations of regulatory or control DNA, which probably includes more DNA than the total concerned in structural genes. The expression of such DNA can often be guessed at, but no experimental approach comparable to the DNA to protein sequence governed by the genetic code has yet been conceived.

The two systems mentioned as exceptions, the brain and the immune system, seem to be much more complex, and there is a third system or process that manifests a still further order of complexity: the process of embryonic development. That is still quite beyond our understanding, and if things go wrong at that stage, the usual result is an early abortion and no significant medical problem arises. The immune system, my own primary field of study, may have as its chief importance the possibility of throwing light on the genetic handling of the central nervous system. Much still remains to be learned of the immune system, but probably even neurologists would admit that immunologists have at least a much clearer provisional picture of the genetic processes at work in their chosen field than neurologists do in theirs. Scientists interested in the brain may be completely sceptical as to whether what has been learnt about the genetics of the immune system will have any relevance at all to the nervous system. Nevertheless, there are some interesting resemblances between the two systems; we speak probably legitimately of immunological memory, for instance. We know that in the immune system, where much more diversity of proteins and flexibility of function is required than in any of the routine systems, ways which we very nearly understand were devised by nature to fulfil these needs. We can feel certain that the even greater needs of the central nervous system for genetic control of its construction in embryonic life and the establishment and maintenance of its memories, skills, and programmes in postnatal life must have been met by something at least as complex as is seen in the immune system. No one could ever provide a clear, simple account of the genetic control of the central nervous system and its bearing on human thought and behaviour. Nor could all the geneticists and neurologists in the world produce an encyclopaedic technical account of such matters. This does not however preclude us from reaching some provisional conclusions that may be helpful in understanding some aspects of human behaviour. If the evidence taken directly from data on the inheritance of human and animal behaviour patterns indicates a large genetic contribution to behaviour, there is no theoretical difficulty in thinking of a genetic and molecular basis of that contribution.

The genetic infrastructure to behaviour

Perhaps the most important of all current controversies about the human situation is concerned with the extent to which human anti-social behaviour is genetically determined. One can emphasize immediately that no form of human behaviour after the cry of the newborn is wholly determined by the genes. Every aspect of childhood or adult behaviour is based on a genetically determined nervous system, a determination that covers its principles of functioning as well as its gross and microscopic structure, but the specific details of behaviour are wholly dependent on the changing environment. The actions of a tennis player are in a sense determined solely by the movement of the ball as it comes across the net from his opponent. The largely unconscious elements of those actions, however, depend on the existence of an immensely complicated apparatus laid down genetically by which visual sensory cues from the ball are converted within fractions of a second into 'a lovely passing drive that raised the chalk on the side line'. The neuromuscular co-ordination that makes a great player is built up on the genetic basis by years of practice and strong motivation to excel. Clearly, what is given genetically must be tempered and honed to become the exquisite mechanism of the champion. Every sports coach, however, is aware that only a very small proportion of boys and girls are potential champions. It is revealing that the current cliché is that the youngster 'took to tennis [or cricket or anything else] like a duck to water'. I shall accept a well-supported but still minority view that in a special sense behaviour is predominantly determined by genetic factors in basically the same way as its genes determine how a recently hatched duckling immediately finds itself at home on its pond.

There are two famous expositors of bird behaviour, Konrad Lorenz and Nikko Tinbergen, who have influenced me deeply. Tinbergen, best known for his studies on gulls but also widely knowledgeable in other fields, has written extensively on the process of learning in birds and mammals. For him, the first need is the provision by the genes of the neural machinery by which the young animal can *learn easily, from invironmental cues, modes of behaviour that will help it to survive*. In man a similar principle must hold, both for qualities of behaviour that are universal, like language, or are characteristic only of small subgroups of people, like capacity to excell in music, mathematics, tennis, or weightlifting. In Tinbergen's own words from his Croonian Lecture in 1972,[1] 'The genetic instructions for the development of behaviour include instructions for phenotypic adaptation ... what is learnt and how it is learnt is prescribed internally within relatively narrow limits. Each different species is programmed for learning in its own way' the adaptations that are necessary. Details of behaviour will often depend on imitation of the parent. Young oyster-catchers will learn from their parents the technique of opening bivalve molluscs by 'hammering' or 'stabbing' although they have the potentiality for both. Chaffinches learn their song by imitation, but they are pre-programmed in favor of learning songs that possess certain characteristics of the standard song of the species.

One of the classical approaches to determine the genetic components of mammalian behaviour is to start with a heterogeneous unselected group of rats and train them for some type of behaviour, such as to find their way through a simple type of maze. Some will be regularly quicker in learning than average, others will take longer than the others and continue to make mistakes. If two lines of breeding are set up, one of rats selected for 'brightness' (that is, rapid learning at each generation), and the other, in which the slowest rats in each generation are used, two results are possible. If the difference between 'bright' and 'dull' rats is not genetic, then each new generation, whether selected for brightness or dullness, should show a similar spectrum from bright to dull amongst its members. Anyone who has ever done an elementary course in genetics will remember Johanssen's beans. In his race of beans, there was a wide range of size but, irrespective of whether the smallest or the largest beans were sown, the progeny showed the same range of sizes. Size in those beans was not an inherited quality. In the case of the 'bright' and the 'dull' rats, however, selection soon gave Robert Tyron, whose experiments are the best known, a strikingly different result. By the time twenty generations had been achieved, there was virtually no overlap; only the brightest of the 'dulls' could sometimes accomplish the maze test in the same time as the dullest of the 'brights'. Genetic factors must therefore be concerned in differentiating rats as bright or dull by this type of test.[2]

Another approach to demonstrating that genes are important in behaviour comes from the study of pure-line strains of mice. Most of the standard lines in the laboratories were developed by cancer research workers between 1910 and 1930 for their own purposes, which had nothing to do with behaviour. Yet people concerned with animal behaviour find quite striking and consistent differences between the different classical strains. One group (Bovet's) found that the strain DBA consistently learned to avoid an electric shock or to thread a maze much more easily than strains C3H or C57B1. Another group (McLearn's) introduced mice to a new world of behaviour by giving them the choice of drinking either pure water of a 10-per-cent solution of alcohol in water. Strains differed sharply in their response: brown DBA mice were almost total abstainers, C3H (an agouti coloured strain) took a little alcohol, whereas the lively, aggressive black C57s got two-thirds of their liquid from the alcohol solution.[3]

There may well be a human lesson here: that accidental genetic factors, which have evolved and survived for some quite unrelated reason, may have a major influence on behaviour toward some new environmental quality. Some patients with the metabolic disease porphyria, a wholly genetic condition, become mentally disturbed, for example, whenever they take barbiturate sleeping pills.

The development of language

At the human level, it has never been fully explained how a two-year-old child can rather suddenly develop the ability not only to make proper use of phrases picked up from parents and siblings, but also to compose for himself new combinations of words which soon begin to show correct grammatical relationships. This will be done in whatever language is used by the child's immediate associates. Here there must be an intense predevelopment of a genetic infrastructure to provide all the still largely unknown qualities for this primary learning of language—everything, that is, except vocabulary, intonation, and grammatical structure of the language itself.

Although Piaget does not discuss the development of human intelligence in infancy and childhood in terms of genetically determined neural infrastructures of behaviour, his approach seems to imply just such a way of looking at what he calls 'internal schemes' which are basic to his concepts. The very regularity with which the mental processes of the child develop demands an intrinsic (i.e., genetic) sequence of changes, becoming progressively ready to be co-ordinated with the requirements of the environment, including social ones. As far as one without much acquaintance with linguistic studies and theories can follow, I am impressed with a recent summary by Delbruck of Piaget's work on the development of language in the child, in which he shows that the symbolic function, private to the child himself, develops long before the effective use of language for communication. The inborn ability of a child or any culture to learn easily any language from its parents and early associates, and of deaf children of deaf-mute parents to acquire a sign language with similar ease, must indicate that the semantic component of language is largely independent of its mapping into actual speech. In some as yet incomprehensible fashion the genes lay down a neural infra-structure that, irrespective of the vocabulary and syntax of the language to be learned, ensures that this will be done easily and result in a semantically effective facility appropriate to social requirements.

Modern biological theory provides no other source for any inborn structure of man than the information laid down in the DNA of the fertilized ovum, provided one bears in mind that the necessary conditions in the mother's body for its development were in their turn constructed in terms of *her* DNA. No matter how far back that regression is taken, no source of information other than linear configurations of DNA seems to be admissible. If one is ever to elucidate Einstein's famous remark, 'The eternally incomprehensible thing about the world is its comprehensibility', one should perhaps add to it, 'to a mind-brain evolved to deal with requirements useful for the survival of average hominids'. I find it inconceivable that the qualities of men like Newton, Laplace, Clerk Maxwell, or Einstein that differentiate them from the rest of humanity were not essentially genetic in character. To express the nature of the in-tuition, the imaginative leap, by which any major intellectual achievement in science or the arts is initiated, in molecular-genetic terms will be the last and greatest task for theoretical biology.

In the present context, however, this interlude is merely another reminder that all human behaviour and its mental concomitants shows unmistakable evidence of genetic control and of evolutionary plasticity.

Socially significant behaviour

The diversity of human behaviour is infinite, and evidence can easily be found to support any moderately reasonable speculation about the origin or significance of behaviour. My present interest is in looking for an understanding of the socially desirable and undesir-able differences in behaviour that we find among members of an otherwise racially homogeneous human population. Influenced, as before, very largely by the ethologists, I have concentrated on three qualities, all of them complex and poorly understood, which seem to be the most relevant in that context. They are intelligence, aggression, and interpersonal bonding or love. In each of these, people differ greatly from one another, and in most of the inter-actions between individuals and groups that give rise to human problems one or more of these three qualities dominate the situation.

Intelligence as tested by standard I.Q. tests is concerned with the differences between people in their ability to handle verbal and other types of problem situations successfully. It is something hard to define but easily recognized. The conventional I.Q. measures are reproducible, and when a group of children showing initially a wide span of difference are retested at a later age the usual finding is that all show an improvement but the relative ranking remains the same. There is a significant correlation between I.Q. level in childhood and adult intelligence tests and success in university education. High distinction in professional and academic life is to a large extent limited to those who by test or by collateral evidence ranked high in I.Q. during childhood and adolescence. Success in other types of occupation, including, according to the literature I have access to, athletics, selling occupations, music, and leadership, is less closely correlated with I.Q. measurements made in childhood.

Aggression is taken as a measure of capacity to drive for a desired end and overcome opposition by other persons in the process. The classical expression of aggression is to gain one's end by killing the opponent. Aggression is a characteristically male quality, not corre-lated with intelligence, and with a variety of socially acceptable as well as unacceptable manifestations. Authority, leadership, prestige, physical courage, and perhaps persistence can be included as well as arrogance, tyranny, cruelty, and many types of crime. In principle, it should be possible to assess and study a dominance quotient (D.Q.) for individuals in somewhat the same fashion as I.Q. Even in the most civilized encounters with another individual, many people find it easy to sense that in that particular situation the other is his superior or vice versa. In many instances, whatever the circumstan-ces, A will consistently sense his superiority to B.

Any human phenomenon nowadays will inevitably be studied in laboratory mice. It is relatively easy to compare aggression in different strains of mice by examining their approach to mice of other strains and their success in fighting. The rank achieved is consistent. Black C57B1 will always win against agouti C3H, and either will overcome A strain (white) mice. Even in mice, however, training and stimulation of confidence by 'arranged' successes can push a mouse above his innate position in the hierarchy of dominance.

Observation and reading indicate a similar, perhaps much greater, liability of status and success in conflict among men, particularly as a result of training—brainwashing?—and experience. Nevertheless, as with intelligence and other human faculties, the necessary infrastructure is laid down genetically and the easily recognizable differences in aggressiveness and related characteristics among individuals subject to the same environmental pressures can safely be ascribed to genetic differences.

Bonding is the ethologists' term for a special relationship between individuals for which such terms as love, loyalty, admiration, respect, friendship, sympathy, care, compassion, and pity may be applied. It covers sexual and parental bonds, and it is probably a legitimate extension to include similar attitudes towards groups of human beings in so far as such groups tend to be personified. Affection toward others is characteristic of female attitudes in much the same way as aggression is of male, but with appropriate shadings both are important in the opposite sex. I know of no way in which interpersonal bonding can be quantitatively assessed or even ranked, yet, just as with aggression, any person of some sensitivity can feel and appreciate when even slight degrees of bonding involve himself. It is a subject of immense interest to ethologists. Lorenz[4] in particular has made a special study of how bonding and aggression interact in geese and normally lead to the triumph ceremony, the highlight of anserine social life.

Those, then, are the three bases of human behaviour from which, I believe, we may most conveniently approach the study of a genetic background to behaviour. I am as aware as anyone else of the bewildering complexity, on the one hand, of neural structure and function, and of the development and manifestations of human behaviour, on the other. Any simplification of that complexity must be a distortion of reality, yet the reality cannot be converted into a working tool that could help in the ordering of human life and human institutions. Brain and mind were evolved *not* to understand their own workings, but to facilitate survival in a world plagued by the antagonism of similar brain-minds in others. If we are to attempt to handle the consequences of the latest phase of our evolution, we must somehow carve something out of reality by which we can get a grip of what is happening in terms that are not too difficult for a person of average intelligence to grasp and manipulate. Expediency in that rather special sense, rather than truth, is what is required. When a situation becomes so complex and significant and effects become so dependent on that complexity that reproducible results appropriate for scientific analysis cannot be obtained, truth ceases to have any meaning. At best, certain stochastic regularities may emerge of little or no value for handling the problems of any individual system.

Intelligence

In a discussion of the matter, R. A. Fisher, a man of nearly the same status in biology as Darwin and Mendel, concluded that intelligence as measured by I.Q. tests was highly heritable, 70 to 80 per cent of the differences between people being of genetic origin, that around 100 genes were involved, and that, in general, high intelligence tended to be dominant over low. In recent years there has been a strange inhibition amongst American intellectuals in the fields of anthropology, education, psychology, and sociology against admitting that genetics played any part in human behaviour or in the observable differences in I.Q. tests. Where low intelligence, poverty, ill health, delinquency, and crime occurred, they depended essentially on cultural factors. This attitude has been traced largely to the development of the cultural history school of American anthropology under Franz Boas, Ruth Benedict, and their followers. Instinctive, that is, genetically based, behaviour was regarded as being shown only in the immediate capacity of the newborn infant to breathe, to cry, and to suck. Everything else was learned according to the cultural patterns of the community into which the child was born. The human species differed from all other mammals in its capacity to transmit abilities and knowledge by language and other types of communication to the next generation. Genetics remained vital for the transmission of physical characteristics, but its importance for animal behaviour had vanished when brain and mind became fully human. Cultural transfer of knowledge and skills had taken over from genetics as the means of evolutionary progress. Culture was presumed to be wholly responsible for the behavioural characteristics of different primitive peoples. It had even affected their mode of thought, giving each culture a unique orientation toward reality, which determined how the minds of its members worked. Essentially, it is this attitude from which the ideology of disadvantaged groups and their supporters throughout the Western democratic-capitalist societies has been developed. The attitude is particularly evident in America in relation to the black ghetto problem, and there is still a sharp difference of opinion between those like Jensen and Shockley, who ascribe the lower average I.Q. in blacks to genetic factors, and a large articulate minority, who would base it wholly on environmental disadvantages. Differences within the academic community itself seem to be lessening, and the recent seminars on human diversity,[5] sponsored by the American Academy of Arts and Sciences, have produced a well-balanced account of the genetic aspects of human behaviour and helped define the areas where differences of opinion are still legitimate. The report of the seminars has been most helpful in revising this and some of the other chapters of the present work.

Jensen in 1973 summed up the position as follows: 'All the major fields would seem to be comprehended quite well by the hypothesis that something between one half and three quarters of the average I.Q. difference between American negroes and whites is attributable to genetic factors.' Most academics in America seem to take the attitude that this may well be true but that there are good social reasons for not adopting a point of view which provides no hope of ameliorating the racial situation. The current attitude against genetics is an obsession that has very much the same quality that Lysenkoism held in the U S S R during the Stalin period, as an attemp to supersede scientific study and understanding by a wholly political concept. As I have suggested earlier, it probably represents a modern twist of attitudes laid down in the last million years. It may be extremely difficult to change, and represents a very important component of the antiscientific attitude of present Western society.

The importance of genetic factors in determining human ability, termperament, and intelligence has been recognized by intelligent people from time immemorial. It was taken for granted in the late nineteenth and early twentieth centuries. Darwin in his autobiography wrote: 'I am inclined to agree with Francis Galton in believing that education and environment produce only a small effect on the mind of anyone and that most of our qualities are innate'. He even suggested that his propensity to collect beetles, since it affected no other of his family or associates, must have been innate! As one who in his youth was affected by the same anomaly of behaviour under similar circumstances, I sympathize with his view that it was determined not too remotely by some uncommon combination of the genes.

With the discovery of the significance of identical twins and of the twin method of assessing the relative contribution of nature and nurture, inheritance and environment, for any measurable quality by which human beings differ, it seemed to be triumphantly established that nearly all the differences that could not be immediately and unequivocally ascribed to injury, infection, malnutrition, or hardship of extreme degree were to a large extent of genetic origin. It emerged too that many conditions formerly ascribed to an easily definable cause—tuberculosis, for instance—were also largely determined by genetic factors. There have been many studies and much controversy about the intelligence and other personality factors of twins. It is usually stated that comparison of identical and non-identical twins indicates that 70 to 80 per cent of the variance of I.Q. in standard middle-class Caucasian populations is of genetic origin and there is no evidence that I have seen to suggest that this finding does not hold for any other racial mixture.

A reasonable case has been made for assuming that part of the correlation with inheritance can be traced to the fact that the home environment is largely determined by parental behaviour which in its turn will be determined by genes of the two parents that in halved and re-sorted collections gave rise to the genes of the children. Even so, no one has been able to claim that less than 60 per cent of the variance is of genetic origin. I suspect that most people, when they look at the differences in achievement and personality of people who they know had much the same social background and education as themselves, would agree with Lord Melbourne that only the intelligent are susceptible to real education, and surely they always educate themselves.

Aggression and bonding

There are no twin studies that I know of on aggression or capacity for love, but there are on criminality, which is likely to include an important component of aggression, and on schizophrenia, which in some ways represents a pathological failure of normal interpersonal bonding. Both show an overwhelming importance of genetic factors.

Probably like most people in my age group, I first read of Lange's work on 'crime as destiny' in J. B. S. Haldane's collection of essays, *The Inequality of Man*, published in 1930. Lange was a German psychiatrist who studied thirty pairs of twins, seventeen sets of two-egg (fraternal) twins and thirteen of monozygotic (identical) twins. They were chosen for study on the basis that one twin, the index twin, had been imprisoned for a criminal offence. The corresponding 'co-twins' were located and their social and psychiatric histories were compared. Of the thirteen identical pairs, the co-twin had also been imprisoned in ten instances (77-per-cent concordance), while among seventeen fraternal pairs only two co-twins had been imprisoned (13-per-cent concordance). Among the concordant identical twins there were several instances showing an extraordinary resemblance in the general life history and type of crime. Haldane, a famous geneticist and at that time politically of the extreme left, concurred with Lange's interpretation. Several similar studies made subsequent to Lange's confirmed his findings, although most workers found a higher concordance of non-identical twin pairs than he did. Eysenck gives combined figures for something over 100 of each type of twin pair, showing 71-per-cent concordance for identical twins and 34-per-cent for fraternal twins.

A more extensive discussion of genetic aspects of aggression, political power, and crime is attempted in the next two chapters. Here I am only concerned to show that there is strong prima facie evidence for a dominating importance of genetic factors in socially deviant behaviour.

Schizophrenia has never been easy to understand. Even among psychiatrists there are serious differences of opinion as to the criteria of diagnosis and as to whether three or more distinct conditions are being lumped together under the one label. The origin of schizophrenia is even more controversial. The commonest opinion among academically oriented psychiatrists and neurologists is that schizophrenia is almost wholly genetic in origin, although the age at which breakdown occurs is probably determined by social circumstances. An almost diametrically opposite position, often taken by psychotherapists and social workers, is that schizophrenia is a result of faulty parental upbringing and without genetic components. A third point of view, becoming increasingly prevalent, is that schizophrenia does not exist. Seymour Kety,[6] who has been responsible for much of the best work on the inheritance of schizophrenia, writes about this view as follows: 'According to its protagonists, "schizophrenia" is simply a label applied by the psychiatric establishment to behaviour which is unorthodox or of which it disapproves but which is actually the result of societal inequities. The social consequences of this third view have been disastrous in terms of loss of interest in mental disease, condemnation of methods of treatment, and a tendency to discredit research in psychobiology.'

It is accepted by all that there is a strong family tendency to schizophrenia, but the heretical third group regard this as merely one of the arguments in favour of the cultural disadvantage hypothesis. To investigate this possibility, Kety and his colleagues undertook a large-scale examination of children taken for adoption in Denmark, where comprehensive records of their subsequent medical history are kept. This is such an enlightening investigation that it seems worth recounting at some length. Some 5500 adoptees were studied, of whom approximately 500 were subsequently admitted to a mental hospital on one or more occasions. For 33 of these, the diagnosis of schizophrenia was regarded as unequivocal and they became the 'index cases'. A 'control' series of 33 was then chosen from the 5000 who never showed mental illness with all the usual statistical precautions to produce a strictly comparable group. Any adopted child can be thought of as having two sets of first-degree relatives—those biologically related and those obtained by adoption. Genetic effects will be relevant to the first set, environmental ones to the second. In the Danish investigations, after much effort 512 relatives of the 66 individuals under study were identified and interviewed. These included father, mother, siblings, and half-siblings, exactly similar procedures being applied to both sets of relatives.

The essential findings were that the biological relatives of the index cases showed 13 cases of schizophrenia, those of the controls 3. Adoptive relatives showed a frequency of about 1 per cent irrespective of whether they were associated with index or control cases. Finally, the same excess of schizophrenic relatives was found for paternal half-siblings of the index cases as compared with the corresponding relatives of the control group.

These results show the direct genetic association and confirm the extensive work on schizophrenia inheritance obtained from twin studies. When 'typical' schizophrenia is present in the index twin, the co-twin is diagnosed as schizophrenic in from 40 to 87 per cent, the differences apparently depending on the criteria for diagnosis. One investigator (Heston) found 46 per cent concordant, 41 per cent with schizoid personality, and the remaining 13 per cent near normal. Gottesman and Shields[7] in England concluded, from an extensive twin study, that specific genetic factors clearly underlie schizophrenia. Environmental factors are non-specific and idiosyncratic. Family studies discussed by Heston[8] also suggest that the basic schizoid genetic factor or factors is modifiable by other genetic traits.

In addition to schizophrenia, manic depressive insanity has strong genetic components, and recent work suggests, but by no means proves, that the genetic abnormality may be on the X chromosome. Senile dementia, and at least two diseases in which dementia at a relatively early age is conspicuous, are also genetic in origin. These have been briefly discussed in the chapter on age-associated disease.

This sketch of the importance of inheritance in most of the significant psychoses reinforces the impression derived from all the studies of normal behaviour in animals and of intelligence and criminality in human beings. Inheritance is a preponderant determinant of all forms of human behaviour that are of social significance. If behaviour depends on the architecture and functioning of the central nervous system, it could hardly be otherwise.

12

Genetics of Power

Power over other people has been central to all history and is the root of human evil. It would be premature to attempt a biological definition of evil, but in some form or other, one must eventually be formulated and accepted. A tentative wording that may have some virtue in stimulating saner attitudes to social difficulties could be that 'to do evil is to exercise power to destroy or diminish the opportunity of a biologically satisfying existence for one or more human beings'. The practical application of such a definition would need interminable discussion of the nature of a biologically satisfying existence, and wherever large-scale action is involved the qualification would need to be added that action harmful to some may not be evil if it brings a concomitant improvement of their terms of existence for a larger number of other human beings. In terms of such a definition there have only been rare episodes in time and space where government has not been by evil men.

Endurance of life is at best a muted conflict against the all-pervading pressures of power; at the worst it is humiliation, torture, and untimely death at the hands of evil men. The problem of pain and evil was evident to the first philosophers and must have been appreciated for generations before men became capable of writing. In the last chapter I have followed the ethologists' lead in arguing that human behaviour is based on a genetically determined neural infrastructure that evolved essentially to its present form in the long hunter-gatherer phase of human prehistory. The basis used for that discussion was that socially significant human behaviour could be looked at in relation to intelligence, to aggression, and to interpersonal bonding or love, and the same three factors must be central to a biological discussion of power and evil.

I am as aware as anyone of the gross oversimplification involved

164

in such a presentation, just as I am of the inadequacy of everything I have written on any biological topic. On every second page of this book is some qualification that biological reality is more complex than I have been able to express; that realization, I believe, is or should be uniquely characteristic of the present stage of biological thought. If science is to be used for human purposes, biology at least, and that includes *inter alia* human behaviour, must be reduced to a selection of facts and concepts that are within the mental reach of those who will use them in human affairs. If the concepts that emerge can cover the significant facts and offer a basis for reasonable contemporary action, this is sufficient justification for their use. I have had enough experience of biology to know that to read a new simple concept can sometimes provide a sudden enlightenment and gratification and, before that has faded, an urge to twist the concept a little to give it greater usefulness. Science can no longer seek ultimate truth but only such aspects of truth as our brains and minds are fitted by their evolutionary history to handle.

Chimpanzees must be the nearest living creatures to the pre-hominid ancestors of man, and the hunter-gatherer style of life seen in full development by the pre-European Australians must clearly have developed directly from something very similar to what Jane Goodall describes for her chimpanzees. They seek whatever edible titbits they can find in their environment just as indigenous children do in Papua New Guinea and no doubt in a thousand other regions at a similar stage of development. Food is not only vegetarian. Eggs and nestlings or termites picked out of a breached mound on a straw will serve just as well, and, when opportunity offers, young monkeys may be caught, killed and eaten. Already with the chimpanzees one can see how the stimulus and the problems of finding *all sorts* of food would be favoured by, and hence stimulate, the evolution of increasing capacity for invention. At times the apes make primitive tools and occasionally one has been seen to throw a stone.

Weapons and warfare

Moving over to what is known of Australopithecus and the other early hominids of south and east Africa, and what can be presumed about their behavioural evolution, the obvious first new development must have been the discovery of weapons which could allow larger animals to be hunted for food. A minimum of fabrication is needed to produce a spear from a trimmed and fire-pointed sapling, or clubs from long bones of the larger mammals, or hand axes from pieces of shattered rock. Once the idea of using artifacts as hunting weapons had arisen, the way was open for their improvement. There was a great bonus for survival with each improvement in hunting technique and as a corollary an advantage to any group amongst whom individuals of better than average inventiveness or intelligence arose.

Along with the development of hunting weapons came the potentiality of lethal intraspecific conflict, murder, and war. Male mammals are always liable to quarrel and fight over mates or status and privilege within the group, but only a tiny proportion of the fights end in death for one of the combatants. When two reasonably matched men without weapons fight, superiority can usually be established and defeat accepted without killing. With weapons available, anger became much more likely to be lethal. Once Cain had slain Abel, war became inevitable and history began.

Within the standard group of something under a hundred individuals all known to each other on a face-to-face basis, growing up within the group and developing the bonds of kinship, killing would almost automatically be recognized as intolerable and under a taboo only to be broken under intense provocation. No such inhibition would be developed against alien groups and everything suggests that in any area where food was relatively abundant and groups likely to interact, perpetual low-grade hostility between groups with occasional acute and murderous contests became the rule. Peking man (*Homo erectus*) was a cannibal half a million years ago, and there is nothing in the mainstream of human history to suggest any significant break in the use of killing as the ultimate sanction of power down to our own times.

What particularly concerns me is the probable genetic significance of this development. Success in war has always demanded certain virtues—bodily strength and skill, courage and intelligence, and ability to hold the loyalty of companions—and from the days of Homer these have provided the prototype of what should be admired in the young male. The genetic basis of such an attitude still persists, however firmly its expression is inhibited by the conventional liberal thinker.

Most realists will probably accept the view that war and lesser struggles between man for dominance form the basis of history and politics, but it is almost unheard of to suggest that the whole complex web of war, leadership, class differentiation, and the admired male characteristics has a strong genetic basis. In his book *Genetics and Man*[1] Darlington says much that seems to bring a proper relationship between genetic and environmental factors in the shaping of human behaviour and is highly relevant to the present thesis. He emphasizes that 'instinct' is not a useful word in discussing behaviour in man. All manifestations of what is called instinct in animal include a large, necessary component based on appreciation of the environmental circumstances in which the instinctive action is to be carried out. In man, the genetically determined neural infrastructure provides the motive force, the drive, but details of action depend on a multitude of environmental circumstances and on the experience and memory of the individual. This holds especially in regard to the qualities related to aggression.

Men in groups

From the beginning of the practice of deliberate homicide between groups, the interactions of aggression and intelligence must have been of special importance for survival and for human evolution. For success in the survival of a human group, improvement in weapons, skill and courage in using them, and strategy and cunning in choosing the circumstances for battle are all almost equally important. But the necessary qualities, at least by contemporary experience, will rarely be combined in one individual. This leads to an important evolutionary question which, like several others, was first asked by Francis Galton in regard to leadership amongst gregarious animals such as the African buffalo and semi-domesticated oxen. He felt that there must be a distribution of genetic qualities basic to leadership throughout the species such that only a small percentage could emerge as dominant males but that such a proportion should be almost regularly present in any viable group of animals. In human terms, the question would become one of asking whether to any significant extent each new human generation shows a *distribution* of socially necessary qualities amongst individuals. One has in mind something that is almost a more flexible, more human analogue of the castes of social insects: queen, male, worker, soldier. The male/female ratio of human births is around 1.05 to 1.12 (i.e. between 105 and 112 boys are born per 100 girls), and is still higher at conception. This seems a reasonable evolutionary development to counter the lower viability of the male, though it is far from easy to see how the initial ratio of around 1.20 at conception is arranged for. At least it provides some justification for thinking that if there was an evolutionary need for a certain distribution of capacities of leadership as against loyal subordination or of high intelligence as against mediocre mental capacity, nature might be able to arrange those too.

Whatever the genetic mechanism that makes it possible, it is a fact that in virtually every sizable human population the distribution of measurable qualities bearing on behaviour will in every instance approximate to the bell-shaped curve of a probability distribution. In every cohort of a million births, one may expect that, given the appropriate opportunity, there will emerge two or three brilliant mathematicians, a dozen scientists of repute, and a national champion in each of a few branches of athletics. What is to be found making up the other spreading flange of the bell curve does not bear thinking about. If we wait until the half million males are eighteen to twenty years old, and assess in any way that may become available their basic aggressiveness, once again the same type of distribution would almost certainly be seen. Capacity for love and loyalty, the third genetic basis of behaviour in our scheme, would be even more difficult to assess, but the results, if we had them, would take the same general form. The three hypothetical basic qualities may be wrongly chosen; each may represent a rather naive lumping together of a dozen more or less distinctive but overlapping and soft-edged aspects of the neural infrastructure. But whatever qualities are chosen, the same sort of distribution would be found and the same sort of genetic mechanism to account for it would have to be sought.

A commonsense evolutionary explanation would be to say that in the hunter-gatherer period a distribution of basic behavioural traits of this sort provided the best average chance of survival in the tribe of around a hundred individuals, all of whom would know each other as persons. What was best for the survival of a hunting tribe or a primitive village is not necessarily appropriate for a massive industrial civilization where communication facilities make it possible to sift out and identify any category of those with special skills or outstanding qualities of leadership or to bring together people of tiny minorities with some unusual combination or intensity of qualities which the majority of the community will regard as psychopathic and dangerous to society. In any large city there will be a host of groups which could never come into being a village, including criminal gangs and, often, cells of political terrorists.

The other virtue of association face to face with a small 'natural' group of village or tribe is that it gives each individual a knowledge, and for practical purposes an understanding, of the whole cross-section of human functions. The almost inevitable result in the city is that the minorities, whether 'good' or 'bad', tend to develop an effective association with only a very small cross-section of the whole community and are not susceptible to traditional values. Acceptance of authority on the one hand and restrained and reasonable use of authority on the other are most likely to work to a community's advantage when those who lead and those who are led know each other as individuals.

This principle of the probability distribution of human qualities and the effect of free communication within large urban populations' in allowing aggregation of individuals with some extreme behavioural characteristic in common allows a useful understanding of how persistence of values generated in the hunter-gatherer phase of human evolution may be inappropriate to modern civilization. Three current difficulties may be briefly analysed from this angle.

1. War in a scientific age

I have given the reasons for believing that the most important behavioural human characteristic for the recent evolution and social development of our species has been the development and progressive intensification of intraspecific killing. At no point in the fossil, archaeological, or historical record has there been any significant break in the evolution of more effectively lethal weapons and more effective methods of using them in war. Nor is there any evidence that the basic attitude of young men toward violence has changed significantly. To quote Washburn: 'Throughout history, society has depended on young adult males to hunt, to fight, and to maintain the social order with violence ... the role requires extremely aggressive action, which was socially approved, learned in play, and personally gratifying.'

In view of the professed revulsion of many liberally minded groups against this interpretation of history, it is well to remember Archimedes, Leonardo da Vinci, Galileo, and, in our own time, Einstein, Fermi and Szilard who wrote the letter to President Roosevelt that sparked the Manhattan Project. All of them, for good contemporary reasons, lent their genius for the production of better weapons of war.

Since 1945, we have had to recognize that the application of science to war has produced a biologically intolerable situation. Physics, chemistry, and microbiology can now provide an immense variety of potentially lethal agents and the means of delivering them with extreme accuracy where they can do the greatest harm. At the same time, biochemistry and physiology have become even more effective and sophisticated in devising ways of destroying vital functions—by the nerve gases for instance—than in curing or ameliorating human disease. An even more significant change is that nowadays no more than a remnant is left of the old socially admired courage and skill in handling lethal weapons in personal combat with an antagonist of comparable quality. Weapons have become weapon systems operated by skilled technicians under conditions of extreme personal danger of sudden extinction, but in a situation that has nothing of the emotional quality of primitive war. Full-scale modern war has become biologically absurd. Nuclear weapons are designed to annihilate a million or more people in an instant, and on the fringes of the holocaust to damage and distort the germ-line and somatic DNA of other millions—nearly all of them noncombatants in old-time nomenclature, and many of the ultimate victims still to be conceived and born. Even without the nuclear bombs and all the 'smart' ways of delivering them, chemical and biological weapons of genocidal intensity could have been elaborated by now, and the varied possibilities for assassination by virtually undetectable means by terrorists or agents of hostile governments are quite terrifying. It is a bitter commentary on the way that genetically based attitudes persist that no public leader anywhere has either publicly stated or acted as if he understood the obvious truth that I have expressed in this paragraph, and that every intelligent and literate individual on earth is well aware of.

2. Terrorism as a political weapon

In every country on earth there are young men in whom the basis of aggression is abnormal enough for them to be recruited as terrorists for almost any cause, no matter how trivial or ridiculous it may seem to others. Higher level conflicts, political, religious, industrial, or ideological, have only to reach a certain level of intensity for the psychopaths to form their underground cells and plan their murders. Just as a child between one and two years is ready and equipped to acquire whatever language is spoken by those who care for him, so young males 'between puberty and marriage' are ripe for violence.

Normal society has of course many ways of using public opinion to keep manifestations of aggression within tolerable limits. Many psychopathic and less intelligent individuals, over-endowed with aggression, will drift into some form of conventional criminal activity. It is only when a significant portion of public opinion is developed in support of some minority attitude that the road to terrorism is nowadays always liable to be taken. The whole process has been described a thousand times and has probably been taking place since the first cities arose in the Middle East. The impact of its modern form is more menacing, probably than ever before, for several reasons.

The first depends on ease of communication in all senses. Ideas and attitudes, once they develop an initial 'take off' capacity, can now spread around the world literally within days, and almost as easily to illiterate as to educated individuals. People, too, can move and make contact with others with similar interests and loyalties, or similar prejudices and hates, with greater ease than ever before.

The second can be called the flange of the bell effect. As the number of people within a population who can communicate to some degree with one another increases, the bell-shaped probability curve of distribution of any of the behavioural qualities that are significant for social order or for social disruption enlarges. The thin flange of the bell, where the 'far-out' types are found, thickens and spreads further. All the rare combinations of human genetic potentialities for good and evil, for eccentricity or psychosis, the people who provide the achievement, the colour, and the dangers and the atrocities of the world, are identifiable to each other and can co-operate. For many minorities, the community becomes all those speaking a single language. In my own tiny minority of research scientists in immunology, the relevant community, now that English is the lingua franca of experimental science, is virtually the world. Terrorism is based initially on two or more individuals, with psychopathic intensity of the genetic potentialities of aggression, being stimulated to violence by some real or fancied injustice. If, as is usually the case, there is some significant section of the community with the same sense of injustice, the initial hard core will find little difficulty in drawing in other young males with strong but less extreme potentialities for aggression. My point is simply that when the relevant group size was a thousand or less, the standard range of genetic diversity would only rarely produce one of the 'far-out' types we are considering, and never more than one.

It is wholly in line with this point of view that in the last decade or two a proportion of women have appeared as terrorists. Dominating and criminal women have been known throughout history, and, without knowing how to account for the process by which such qualities appear, we can be certain that much more rarely than in males a proportion of females have the genetic configuration needed for the exercise of dominance in criminal fashion.

The third reason for the menace of terrorism today is the availability of weapons. Any determined group can in one way or another obtain material for bombs and light weapons like automatic pistols and submachine guns. If, as is usually the case, a terrorist group is sponsored secretly as a matter of policy by a 'nation state', more sophisticated weapons will soon become available. Obsession with power is not incompatible with high scientific achievement, and a major part of the menace of terrorism today is the threat of use by irresponsible groups of the new agents for mass killing in the context of political terrorism. Like many others, I strongly oppose extension of nuclear power, and especially the entry of the breeder reactors, not for their danger as such but because of the great increase in the amount of plutonium that will become available for malevolent use by political terrorists, by the sovereign states that are looking for cheap nuclear weapons to bolster the ambitions of some ruling minority and the nuclear middle powers and super powers. Traffic in plutonium, built up by some unholy alliance of paranoia and state policy, could be the trigger that in the end will destroy civilization.

Forth, and finally, scientific and technological progress inevitably creates vulnerability. The larger and more efficient a power generator, the more harm that will be done when it is put out of action. The technique of industrial blackmail by workers in vulnerable industries is moving in such a fashion within just acceptable limits, but there is no doubt that its lessons are being learnt by the potential terrorists on the periphery.

In all these forms of terrorism, we are dealing essentially with manifestations of the same potentiality for aggression which, contained and tempered by the bonds of the individual with the community, were essential to survival in primitive human groups. The necessary workings of genetically based human diversity, the enormous enlargement of the group in which communication is possible, and the minute proportion with whom face-to-face contact and group bonding is possible are responsible for the change. It is a consequence that was probably quite inevitable once circumstances forced the first hominids to devise weapons for hunting which could kill men as well as animals.

3. Abortion and infanticide

Another highly characteristic difficulty of the present time is of a wholly different character but can equally be traced back to the evolution of behavioural potentialities in the hunter-gatherer period. As with every other mammal in a state of nature, the primitive human populations reached by complex natural processes a state of fluctuating equilibrium with significant expansion of numbers only when new circumstances such as entry into a previously unpopulated but humanly inhabitable region made it possible. Survival was always a struggle and, especially in the young, mortality was very high. To maintain a nearly stable population required a birth rate that would be of the order of 40–50 per 1000 per annum, corresponding to about a dozen completed pregnancies for each fertile woman. Having regard to illness, miscarriages, and stillbirths, this meant that an almost unbroken physiological sequence of pregnancy and lactation was the lot of women from puberty to the menopause. For any group to survive, strong interpersonal bonding of all types was necessary within the tribe or village. Children who survived were precious, to be loved and protected; sexual bonds between individuals needed to be firm and lasting, with monogamy the norm. And, as is characteristic of primitive people today, the obligation would be accepted to share any good fortune throughout the group. Loyalty to the group in defensive of offensive violence was axiomatic.

Aggression toward the outsider, love and loyalty toward the in-group were what dominated the values of the hunter-gatherer period, and we have inherited their neural basis. Here we are concerned with one of the areas of modern life where the bonds within the group, instead of aggression toward the stranger, have caused difficulties.

The pleasure of heterosexual activity was nature's only way of ensuring the unbroken reproductive function of a woman's maturity. Subject to the inevitable diversity of intensity and effectiveness of all genetically determined function, that attitude has persisted unchanged into a period where the biological requirements are very different. Probably most healthy young women nowadays, in comfortable surroundings and knowing that childbirth has virtually lost its dangers, would like to have a family of four or five healthy and affectionate children. They know, however, that if they are to enjoy the extra luxuries that others have, two children early in marriage is enough and then both parents must earn money for those luxuries. Many will be aware of the urgent ecological necessity to reduce population growth to zero or below, but I have no belief that such an essentially intellectual understanding would have had any effective force in bringing it about. The convenience of modern contraception and zest for money and enjoyment can take nearly all the credit for the standard two-child family. Keeping up with the Joneses is an aggressive, rather than a bonding attitude.

Accidents will happen, and an inevitable aspect of the two-child family ethic has been to make abortion respectable. There are three ways to achieve birth control, all precisely equivalent in their biological effect and significance. They are contraception (A) by methods which require conscious interference with the sequence of intercourse and (B) by those which do not; abortion which may be either (C) by 'menstrual regulation' at the earliest suspicion of pregnancy or (D) by standard surgical interference at any time in the first three months; and, finally, (E) infanticide immediately after birth. Three of these, A, D, and E, have all been practised widely throughout history but under religious taboo in most communities. Contraception (A) developed middle-class respectability in the late nineteenth century and became essentially universal in all literate communities affluent enough to afford it when the Pill (B) materialized. Abortion (D) went through a phase when it was legal only if some real or nominal medical damage to mother or infant could be foreseen if the pregnancy were allowed to continue. Now public opinion, wherever it has shaken free from religious inhibitions, seems to be almost unanimous in supporting abortion of every pregnancy when the child is not positively desired by the mother, but also in making every effort to remove the old disgrace of being an unmarried mother so that the woman's choice could be a balanced one. Menstrual regulation, which is of course a mere synonym for very early abortion, seems likely to become more and more common. Like many medically trained people, when I think of the Pill I am mildly unhappy at the thought of between twenty and thirty years' almost continuous interference with the complex hormonal sequences of a woman's sexual rhythm. Intuitively I believe that, for intelligent women with considerate partners, an intelligently handled use of the safe-period technique, with the best and safest method of very early abortion as a backup, is the most socially and biologically desirable approach to contraception. But for most couples, the Pill, vasectomy, or tubal ligation will probably always be the most practical solutions.

Infanticide has a long history. In many primitive societies, immediate infanticide of any *abnormal* product of conception, including twins, was mandatory. Before the European explorations, Tahiti and presumably many other island communities had avoided, by isolation, most of the infectious diseases that elsewhere served as a check on population increase. There, infanticide had become biologically unavoidable and a variety of methods and rituals developed around it. In most regions, however, the taboo was much more rigid against infanticide than against abortion, and it continues to be regarded as murder in most societies. I have discussed the related but more controversial problem of the handling of serious genetic abnormalities in Chapter 9.

It is clearly a prime biological necessity that, as part of the drastic switch of hormonal function from the control of pregnancy and the birth process to the initiation of lactation, the mother should develop an intense and immediate interpersonal bond with the newborn infant. No one seeing or experiencing this reaction can doubt the existence of a genetically controlled basis, ready and actively sensitized to respond to the environmental and bodily experience of birth with this explosion of maternal love. It seems very unlikely that infanticide of healthy infants will ever be tolerated in a stable society. With a sudden disintegration of civilization after major nuclear war, it could be another story.

Power attitudes come into the problems of birth control only rather indirectly, but potently. As long as war remains the final sanction by which human disagreements are arbitrated, that one of two antagonistic groups that has the larger number of young men, i.e., of male infants born eighteen to thirty years previously, will have an advantage. There is a convention abroad to write little about this aspect of war and politics, but almost all the religions that take their myths and dogmas seriously oppose any type of population control amongst their devotees. The significance of this is already very evident in Northern Ireland, and one of the more likely and most unhappy results of a majority move to a rational two-child family will be a deliberate use of religious taboos to increase the fighting or voting power of illiterate and suggestible groups.

13

Human Diversity

Individuality may well be the quality that each of us most cherishes; somehow we must be recognizable as different from anyone else, and, by implication, more valuable. Recognition is primarily by facial structure but is always reinforced by characteristics of facial mobility, by individual quality of voice, and idiosyncracies of usage or pronunciation of words; not quite so constantly relevant are stance, gait, or clothes, but one of them may combine with the others to help define the person met as the individual known to us from previous encounters.

Man is not the only animal species with this need and power to recognize a specific individual from every other member of the species. It is a near miracle to most people that a lamb can always find its own mother in a flock of a hundred sheep or a gannet recognize its mate in a nesting colony of thousands. Those examples give the clue to the evolutionary significance of diversity within the species of most or all mammals and birds, and above all of human diversity. It is to allow bonding between the two parents and between parent and young in all those species where care by one or both parents for the young is a necessary part of the reproductive cycle. For animals that provide no parental care or guidance for their offspring, the only biological need is to recognize the species and the sex of the other individual. It may not be accidental that most of the warmblooded vertebrates who develop individual bonding between mates or between parent and offspring nearly always show the development of a hierarchy of dominance in the adult males, and to a lesser degree in female and juvenile groups. Order within any community of gregarious birds or mammals usually demands such a hierarchy, and, if it is to function, each individual must know his place, in other words be able to recognize those individuals who are his superiors and those whom he can dominate. This requirement, which holds for most species of baboons, monkeys, and apes, probably accounts for the limitation of primate troops to something under a hundred individuals. This must also have been the case for the hunter-gatherer groups of the early hominids and of *Homo sapiens* up to the beginnings of the agricultural-urban revolution about 10 000 years ago.

In man, even in his hunter-gatherer phase, there was another requirement for diversity within the group in the possession of certain skills and social responses. One has only to look in imagination at the constitution of an early hunting group to realize this. Assuming it is around a hundred strong, it will contain perhaps thirty adolescent or mature males. Among these must be found a headman who can lead and control the band, and a deputy who can take his place in an emergency. It would favour survival of the group if one or two of its members had special skill and even some originality in weapon-making. For the rest, the requirement is loyalty and obedience, courage in conflict with animal or man, and the self-reliance to handle his own emergencies. Clearly, some such degree of genetically based diversity within each such group would strongly favour its survival and that of its component members. If the generation of such a distribution of capacities in small groups lay within the competence of human genetic mechanism, it would inevitably have emerged in the course of evolution.

I have found no extended discussion of diversity within primitive groups, but De Vore, from long experience with the South African bushmen (now called San), has said that, although anthropometrically the San are rather more homogeneous than urban Americans, one finds the same wide diversity of personal characteristics as anywhere else. Some are dull, some are bright, some are very creative, and so on.

Genetics of diversity

Diversity of human quality must have become of increasing importance with the development of agricultural communities and the rapid development of specialized function in society. At this stage, however, it is best to confine attention to hunter-gatherer man and the other gregarious mammals from which, in these respects, he did not greatly differ, and to look at some almost universal aspects of the problem of diversity within a species. Here we are concerned with requirements at a deeper level than the capacity to recognize animals as known individuals. The whole process of evolution, as has been stated or implied several times, already depends on the accumulation within the gene pool of a wide variety of alleles, mostly recessive but often co-dominant. These are necessary to allow effective occupation of a new ecological niche for which the standard phenotype is inadequate. The ideal arrangement, since the special qualities of a new niche can never be known until it is experienced, is to have in the species population a very large variety of phenotypes expressing some genetic change but very few examples of each. This requirement is so important that special mechanisms to achieve it have been evolved. Darwin made a special study of fertilization in flowering plants and described several methods by which primroses and other plants managed to ensure that fertilization was always by pollen from another plant. Nature seems to abhor inbreeding. Somewhat similar methods to foster cross-fertilization can be found in some colonial marine invertebrates. In higher vertebrates the tendency is to find barriers against interspecies matings, but, except for the incest taboo in man, no specific hindrance to inbreeding. However, when wild mice caught in barns, fields, or houses are examined for blood, cell, and tissue antigens, they show wide heterogeneity and essentially the whole range of antigens as have been studied in the strains of the same species that are maintained in immunological laboratories. A multiplicity of differences in the mammalian qualities recognizable by laboratory tests must almost certainly indicate a similar polymorphism of the qualities by which an individual can recognize directly differences between members of his own species.

For most mammalian species it seems likely that the standard processes of mutation and recombination occur at a high enough rate to provide the required degree of diversity, particularly in species with a large gene pool. Man is presumably no exception. However, simple facial and other visible differences by which persons can be identified represent only a small and relatively unimportant part of human individuality. We are almost always much more aware of the diversity in the mental characteristics of human beings. If the basic qualities of intelligence, aggression, and interpersonal bonding and their derivative socially relevant behaviour are based on a genetically controlled neural infrastructure, then the 'multiple polymorphisms' of various aspects of behaviour must be controlled through these neural configurations. It is a point that will need more discussion at a later stage.

Mating groups

With the development of urban civilizations in the Middle East and the appearance of professional armies, highly skilled artisans, priests, and kings, as well as the anonymous mass of peasants and labourers, a new form of diversity became necessary. It must very soon have become evident to leaders and administrators that when the need arose for people who could design a new building, provide acceptable decoration in colour or sculptured relief, or make good weapons and reliable armour, some of those who volunteered or were compelled to undertake the work were much better suited and rapidly became more adept than others. A small proportion of people in every human population will be found genetically pre-adapted to learn relatively easily any skilled manipulation of a new type that may be called for. This is exactly analogous to what happens when any animal species is confronted with a new ecological niche to occupy. Continuing the analogy, the son of a successful artisan is more likely to marry the daughter of another artisan of similar skills than to go outside the craft. If this continues for generations and the quick learners and more adept executants remain within the craft community while those who are unsuccessful drift to less demanding occupations, it is inevitable that a mating group interested in and genetically fitted to become expert in the particular craft will come into being. This will hold to a significant degree despite the inevitable effect or regression toward the average quality of the original unselected population. Regression is a genetic phenomenon characteristic of qualities derived by polygenic inheritance, i.e., depending on the joint action of many genes, and known to be of special importance in the inheritance of intelligence. In Terman's famous study of gifted children in California, he found that his subjects, chosen on the basis of an I.Q. over 140, married people of good intelligence but with a lower average I.Q. than their own, and had children with an I.Q. distribution well to the upper side of the scale but with an average significantly below 140. Similarly, dull normal people with an I.Q. of 70 will have children whose average has moved up toward the mean value of 100. For a mating group with high capacity for some specific skill to persist, there must be a selective process to eliminate those who regress, and to take in from outside recruits of proven capacity. Even if skill were wholly dependent on cultural factors, this would represent an evolutionary social process. If we are influenced by Tinbergen and other modern scholars, or even put weight on ordinary commonsense and traditional wisdom to accept the important role of innate genetically controlled capacity to learn, the biological quality of this emergence of groups with special skills is immediately evident.

Human diversity might almost be compared to the distribution of metal ores in the earth's crust. Through 90 per cent of the igneous rocks all the elements are distributed almost as uniformly as they probably were when the earth took shape 4500 million years ago.

But almost from the beginning, processes potentially capable of diversifying that distribution have been active: melting and fractional recrystallization, solution in hot brines with subsequent differential absorption or crystallization of the various components. Even biological processes can produce geological concentrations of certain elements, notably iron, calcium, and phosphorous. In ways of this sort, ore bodies are laid down and in their turn are subject to similar processes and others associated with erosion, sedimentation, and selective sorting of particulate material. So we have the existent (and rapidly disappearing) concentrations of rich ore in relatively tiny pockets and much larger regions of progressively poorer ore where concentration has been slight or formerly concentrated or has been spread by erosion and water transport over wide areas. And of course the process still goes on, as in the regions of hot metal-rich brine in the deeps of the Red Sea where Arabia and Africa are slowly rifting apart.

The special abilities of mankind in science, literature, art, music, fine craftsmanship, architecture, and engineering are the precious and semi-precious metals that give sparkle to humanity. From hundreds of minor mutations scattered through the gene pool the orderly 'random access' and recombination, followed by selection and concentration of capacities in loosely structured but still very definite mating groups, has given us the diversity of talent which is the real glory of mankind.

Range of diversity in man

Before elaborating on the social and behavioural aspects of human diversity, it is probably desirable to say a little more (see pages 101-5) about the *range* of diversity in man. As an immunologist, I can begin first with what we call the major histocompatibility antigens, which in complicated ways decide immune reactions against kidney grafts and the like. These antigens, the HLA series in man, have been mentioned previously in relation to the occurrence of autoimmune disease (page 97). They are obviously important for clinical medicine, but we really do not know what their function is in the normal physiology of the body. They are of immense interest to immunologists like myself because they are, as it were, the markers that tell where the cells came from. It is only simplifying the picture a little to say that a surgeon can use any part of his patient's body to repair damage because the antigen markers on the transplanted cells certify that they are of an acceptable type, while with only one exception—an identical twin—tissue transplanted from anyone else will be rapidly rejected, because it has the wrong markers. Plastic surgery, however, is a twentieth-century discovery, and the histo-compatibility antigens must have developed in the course of evolution for some other reason. There are several theories, but none is firm enough, or even interesting enough, to elaborate here. We must simply accept their existence and say what we know about them.

In man, we find at least thirty clearly distinguishable types of these complex protein antigens—the exact number is not yet known—four of which, selected almost at random, are present on any cell of a given individual's body. The numbers of ways in which thirty (say) different antigens can be arranged in groups of four is approximately 4 million but the number of possible antibodies that can be produced by all members of the species may be no less than 100 million, according to Jerne. Both figures indicate the immense degree of diversity that can be arranged if the evolutionary requirements call for it. Apart from the brain, where diversity of structure and function as between individuals is much greater and much more significant than in the immune system, diversity in other organs and tissues is relatively, but only relatively, small. A few examples of these human polymorphisms, as they are technically called, which may have some general interest, can be mentioned. The number of alternative forms of the enzyme G6PD—mentioned earlier, in relation to tests for monoclonality—is at least eighty.[1] Most are very rare, a few are common; they may be associated with no apparent abnormality of function, but some give rise to severe pathological effects. Blood group differences are the most widely studied of all human polymorphisms. Confining ourselves to the two practically important sets, there are in the first place four A B O groups which determine whether or not one person's blood is suitable to transfuse into another person's veins. The Rh blood groups that are concerned with Rh disease of newborn babies have a complex type of inheritance based on eight common arrangements of three genetic units: these arrangements, taken two at a time, one on each allele, give sixty-four possible combinations, or which fifty-three were said to have been observed in a recent compilation. Whenever really detailed study is applied to one of the blood groups, minor variants persist in turning up, which of course is in line with everything we know about mutation. Close examination with assessment of amino acid sequence and other minutiae will nearly always indicate a certain heterogeneity or polymorphism in most body proteins; this may or may not have some effect on the efficiency of the cells and tissue involved, and if inefficiency is produced, it may or may not produce symptoms.

The list of human polymorphisms of this sort could be prolonged indefinitely. Most of those mentioned have been given close scrutiny because some of their variants are responsible for disease. It may be that others not concerned with disease may on occasion give rise to equally large numbers of alternatives still quite unrecognized. What must be grasped is that all these chemically defined alternatives that I have mentioned represent what critically minded scientists call hard data, facts that can be reproduced readily by others and that admit of only one immediate interpretation. The abnormality in one of the protein chains of haemoglobin, which is responsible for sickle-cell anaemia, is no more than a single amino acid change from the normal GLU to VAL (glutamic acid to valine). This is a universally accepted hard fact. The accepted hypothesis that the change spread through African populations because the heterozygote with only one S gene had a higher resistance to malaria is probably true but has a much 'softer' quality. Hard data, then, can show that in man the effects of very large numbers of mutations can be unequivocally demonstrated. Some are large effects, many are small. The facts provide an obvious justification for assuming that, for a system like the central nervous system, for which we have not yet found, and may never find, a means for detailed genetic study, a similar wide range of mutations is extremely likely to be present. If behavioural data compel us to postulate this, we have firm analogies from which to build up any such hypotheses.

Behavioural diversity

Reverting to the general topic of human diversity as manifested in abilities and attitudes, the qualities that concern society, I have indicated the likelihood that when people's working activities became specialized and required special skills for success, rather well-defined mating groups were likely to emerge. The important factors that define the mating group are language, colour, and social class, accessibility, including locality and ease of movement, common interests, and common place of work. This may well provide a mechanism for the concentration of genetic factors for special skills, including intellectual ones, and at the same time give rise to a continuing emergence of people with new potential skills that could allow them, for example, to carry out skilled assembly routines in a gravity-free space environment. Yet any discussion of such capacities for skill leaves a great range of human diversity untouched.

Many attempts have been made to sort out the socially significant qualities by which men differ, and some of the modern ideas may be mentioned briefly. Differences in intelligence in terms of I.Q. have been dealt with already. According to Denis Gabor,[2] a capacity to live satisfyingly to oneself without serious dissatisfaction to others in society requires a right place in an *ethical quotient* as well as in an I.Q. Sheldon used the words cerebrotonic, somatotonic, and viscerotonic to indicate temperaments corresponding to ectomorph, mesomorph, and endomorph as names for the corresponding body builds. Eysenck[3] analyses personality primarily in terms of intelligence, extroversion-introversion, and neuroticism, i.e., the contrast between the unstable, emotional, neurotic person, and the stable, unemotional, non-neurotic. In this book I have contended that the three basic features important for society are intelligence, aggression, and capacity for bonding, and I have tried to derive them from the nature of human associations in the hunter-gatherer phase. After reading fairly widely in this field of human population studies, certain conclusions seem to be self-evident:

(1)   that every human being is unique;

(2)   that any measurable quality not showing a Mendelian-type inheritance will give a roughly Gaussian distribution; it will however, show less than 100 per cent correlation with any other quantifiable characteristic;

(3)   that investigators tend to find evidence for the importance of qualities thay happen to be interested in and can measure.

It is clearly necessary to make some assessment of how the approaches to human diversity of the three men I have mentioned—Sheldon, Eysenck, and Gabor—can be fitted into the scheme I have adopted. For reasons that may well be inadequate, I feel that Sheldon's types of temperament and Eysenck's stability-neuroticism gradient are real enough but are not of special importance in the functioning of society. I am more impressed with the possibility that Gabor's ethical quotient and Eysenck's introversion-extroversion may refer to a fourth quality that should be included as something more than can be covered by the concepts of aggression and dominance. Essentially, what Gabor is concerned with is, on one side, the application of aggression and leadership to antisocial ends, and, on the other side, to socially admirable action 'beyond the call of duty'. With a slightly different emphasis, the ethical quotient can be looked at as defining behaviour within a loyalty-subversion gradient. On the side of loyalty we have acceptance of majority opinion in most aspects of social behaviour, honesty and truthfulness, maturity, dedication, stability. On the subversion side, cunning, commercial swindling, theft, insolence, insubordination, immaturity, any use of aggression outside the accepted limits. This brings us close to Eysenck's contention that the basic difference between the introvert and the extrovert is the readiness with which they respond to social conditioning. Introverts are more readily educated and conditioned than extroverts; in Eysenck's view, because of their higher level of 'cortical arousal' in the brain. Perhaps social maturity and social immaturity may be the most revealing as terms for the qualities within this complex that are most directly influenced by the genes. It is a necessary characteristic to be developed over the course of adolescence to conform to community attitudes. As always, people differ in their capacity to respond to the requirement. Mature, well-conditioned, and socially responsible people may have most of the virtues, but may equally find themselves unattractively far to the right of the political spectrum. Delinquency and crime will always result from the interaction of genetic and environmental factors. Of the genetic elements, those basic to a high level of aggression, extroversion with poor capacity for social conditioning, with perhaps a separate P component suggested by Eysenck as being characteristic of the 'psychopathic personality', seem to be those that are mainly responsible for the development of criminal behaviour and lesser forms of delinquency. Probably all that can legitimately be said is that those who for genetic reasons fail to develop the normal responses by the normal times will be more susceptible than average individuals to environmental pressures toward criminal or antisocial behaviour.

My picture of human diversity as far as temperament, attitude, and behaviour are concerned, and with relevance only to role in an accord with the community, is drawn at two levels.

At the first level the basic factors are (1) intelligence, (2) aggression, (3) interpersonal bonding or love, (4) social maturity-immaturity or introversion-extroversion. All of these are presumed to be based on polygenic inheritance; each will show a Gaussian distribution of degrees of the quality being measured in any population reasonably uniform in regard to other qualities.

At the second level we have the genetic background for capacity to learn and excel at various special skills, including intellectual ones.

At both levels there will be development of mating groups in which special genetically based capacities for social success of any sort, admired skills, high intelligence, administrative and military capacity, will tend to become concentrated in the fashion first recognized by Francis Galton in his *Hereditary Genius*. It is probable too that mating groups based on qualities that are socially less desirable also exist and in some regions have maintained themselves for centuries. The Thugs of Central India and other bandit communities of the past, criminal groups in modern cities, and nowadays activist, terrorist groups in many countries, have all had their own mating groups with inherited as well as cultural qualities being needed for their development.

In all these established or incipient mating groups there must be a continuous slow change in the number and characteristics of their members with a constant leaking away of families and individuals from change of occupation or of abode. With the development of a vast pool of essentially unskilled labour, which includes those doing 'skilled' work that can be readily taught to people within the central part of the distributions for intelligence and capacity to accept training, an extremely large anonymous pool is developing in which mating groups are almost non-existent and marriage mostly determined by proximity and wage or salary level. It seems likely that in a modern Western community the process of diversification in the main mass of people is rapidly vanishing.

This leads to the discussion by Denis Gabor in *The Mature Society* of what is to be done about the vast number of people with no special skills and middle-level or low intelligence. Most people in the workforce today would become redundant if the world suddenly became logical and decided to produce only what was biologically necessary for health and comfort in living. One has only to think of the standard naive Utopia, with war and weapons-making abolished except as needed for the maintenance of law and order, with due regard for the long-term carrying capacity of the planet and with families averaging 1·7 children to allow a slow drift down in population to a stabilized level of about 2000 million. In such a dream world we could also postulate the use of the best modern techniques to produce all that was needed to maintain food, clothing, shelter, and education for the whole human species. That could be done probably by 20 per cent or less of the available workforce. Even today, a great proportion of the work by which people earn their wages is completely unnecessary in the sense of being unrelated to the production and distribution of the necessities and comforts of life. In Gabor's opinion, the three great problems of the immediate future are the elimination of war, the control of population, and the use of leisure. The last may become the most difficult to solve because leisure means unemployment. If only those were employed who could offer profitable skills, unemployment would fall almost wholly on those of mediocre or low intelligence and manipulative ability. Somehow the possibility of a tolerable existence must be made available to those who have superficially nothing to offer society. This is in essence the objective of modern left-wing political activism and it is as worthy an objective as any other—unless one happens to be a geneticist with an overriding concern for the health, intelligence, and co-operation of future generations. Even with those prejudices, I am sufficiently doubtful about the possibility of maintaining or improving human quality in the long term by any eugenically oriented social measures, to wonder equally whether what Galton would have regarded as subsidizing the unfit will have any effect in leading future human evolution in what he and I would regard as the wrong direction. The future is too full of uncertainties to be confident about anything.

To say that some way of occupying leisure must be found for 80 per cent of people without regard to their industrial productivity is likely to be offensive to the majority of people and politically is wholly unrealistic. The social and moral necessity of regular work for all male and female adults who are physically capable of it is still the rule in every existing society. The position that has been developing progressively since the beginning of the century has been met in various ways: by reduction of working hours, extension of fulltime education to the age of eighteen or higher, and the development of new forms of secondary and tertiary industries. In one way or another, it will always be politically necessary in any democracy to provide the necessities and decencies of life to everyone without too much concern for the value of his contribution to the pool of goods and services.

It may be more important to consider what may be the long-term effects of such a human ecological situation as is developing now. We have postulated in effect that mating groups will continue to develop around those who are successful in occupations where rare or relatively rare capacities are required. This will give rise to a substantial number of relatively small groups of high potential social value, both for existing and future skilled avocations. They will be in high demand, and if the children born within such groups include a high proportion with potentiality for specific skill to maintain the size of the mating group against the counteracting process of regression toward the mean, a genetically fixed capacity for some distinctive skill might become statistically predictable. One would expect, too, that cross-mating between distinct groups of this type would also give some interesting and successful new combinations. Some of the genes would drift away by cross-mating with people of the undistinguished middle, but under normal circumstances would give rise to very few interesting combinations in that milieu.

The possibility seems there for an increasing dichotomy into two sets of mating groups that are relatively open within the set but allow very little mating across the boundary between sets. Whether this would ever lead to a division or at least an incipient division into two species may be doubted, but if society took the form of a justly and efficiently controlled meritocracy, there might soon be a diminishing movement in either direction, and if this continued for some hundreds of thousands of years a split would no doubt occur. But it is not realistic to believe that any human pattern of social organization could remain stable for 10 000 years, let alone twenty times that span. It is not likely that Wells' time traveller would 'really' have found two species of man in the distant future, but I believe he could well have found an even greater diversity within the species than we have now.

Is a specific genetic factor involved in the evolution of human diversity?

The more one ponders over human diversity, the more it seems to come into focus as one of the three great specifically human qualities, the other two being language and intelligent conceptual thought. Since the socially significant features of human diversity are wholly within the broad sphere of behaviour, all three of these distinguishing qualities must depend on the genetic control of the structure and functioning of the nervous system. It would almost seem worth considering whether some common factor was concerned in the evolution of this whole complex of activities on which human behaviour is based. One senses a specifically enhanced genetic lability in some function concerned with mutation and recombination in the genome as it involves nervous system development. Any sophisticated biologist will smile cynically when I suggest that some quality in DNA-handling enzymes could be involved.

Nevertheless, if there has been active genetic modification during the last million years, what other immediate agency can be suggested? To elaborate a possible hypothesis in the most general terms may be unjustified, but in the absence of any alternative it could conceivably help to direct research in ways from which specific tenable hypotheses might emerge.

As a preliminary to any positive suggestion, we may look back at some of the evidence that modes of handling DNA, presumably by associated enzymes, exist in addition to those concerned with replication and simple repair. In the first instance there are at least three different mechanisms by which ultraviolet damage to DNA can be repaired in *E. coli*, and for the most complex, 'post-replication repair', an elaborate series of changes must be envisaged which probably involve enzyme functions not yet clarified. Then there is the classical problem of how the virtually unlimited number of antibodies that any individual can produce come into being. Almost the only point on which all authorities are agreed is that the process is a genetic one: the configuration of each antibody molecule is laid down by genes in the cell that produces it. In one way or another, an enormous number of slightly different DNAs are developed, and, for this to happen, some quite elaborate ways of handling DNA must be postulated. My own favourite suggestion is that of Baltimore: that there is a special mechanism by which one set of enzymes removes short lengths of DNA, starting at three or four defined points in a certain segment and then another set refills the gap without regard to the proper sequence of nucleotides by using what can be regarded as a non-specific, i.e., completely error-prone, enzyme or complex of enzymes. Something more or less similar might be needed if large amounts of genetically based diversity must be built into human brains and minds.

A totally different type of genetic phenomenon is represented by the random repression of an X chromosome in all cells during the embryonic development of female placental mammals. This is the Mary Lyon phenomenon, already referred to on page 106, in regard to tests for monoclonality in tumours or other collections of cells. How such a type of random choice between two similar X chromosomes is made is unknown. In yet another area we find insistent hints that there are some inexplicable regularities in the types of somatic mutations, including tumours, produced by physical and chemical mutagens.

As I have discussed in Chapter 5 and more fully in a previous book, the types and combinations of genetically based degenerative disease of the central nervous system suggest that a number of different DNA-handling components must be involved if the various clinical pictures are to be interpreted. It is clear that much is still to be learned about the number and genetic control of DNA-handling enzymes if we are to account for these phenomena and others that will undoubtedly be discovered in the future. If our basic hypothesis of ageing is approximately correct, it would be equally legitimate to think of a series of changes (controlled by structural genes) in DNA-handling enzymes which increase the frequency of errors of a particular type in genes controlling the course of embryonic and immediate postnatal development of the central nervous system. In all probability the genes that direct the morphogenesis and primary circuitry of the central nervous system are in the form of regulatory or control DNA, concerned not with defining the structure of a protein gene product but with the on and off control of structural gene activity.

One might postulate a situation with some formal similarity to that concerned with the generation of antibody diversity, particularly as involving both germ-line and somatic processes. What one pictures at the germ-line level is the occurrence in the evolving hominids of more genetic errors in the replication of control DNA concerned with central nervous system development than in structural DNA or in other types of control DNA. Most of the errors will concern the timing of gene activity and are individually rather trivial. One can think of, as examples, a few more or a few less cells produced in a certain phase of development or a change in the potential number of synaptosomes in a certain collection of neurons. Particularly during postnatal development one must assume a continuous process of somatic self-moulding and adaptive change by which what is given genetically is modified by sensory input to build up an adequately functional central nervous system. In such a developmental process quite minor genetic changes of the type I have suggested could make learning of some special skill easier or more difficult. It must remain an open question whether similar processes could also affect the equivalent control DNA in somatic cells. Without attempting to suggest in detail how this could happen, one can sense that such a capacity could help to provide the diversity and flexibility of human behaviour. Any behavioural qualities that can be ascribed to genetic error in the germ line will persist according to the standard rules, although on the hypothesis we are developing any further modification by somatic genetic processes will probably not be expressed in descendants.

Such an hypothesis could provide an interpretation of the outstanding aspect of human evolution, the doubling of brain size in perhaps not more than a million years. In very round figures, *Australopithecus*, who flourished in Africa about 1·7 million years ago, had a brain occupying about 500 cc; *Homo erectus*, around 500 000 years, is recorded as 800 to 1100 cc; while the general opinion is that *Homo sapiens*, with average cerebral capacity of 1350 cc, probably developed rapidly from 250 000 to 100 000 B.P. (Before Present). Another point in which modern man differs from anthropoid apes, and therefore presumably from his hominid ancestors, is in the greater number of neural pathways and connections associated with increased numbers of synapses per neuron. Quantitative increase in cells and their connections could be the simplest manifestation of a modified error-proneness on the reasonable assumption that an increase in anatomical facilities will be made use of, initially by adaptation but with casual suitabilities being exploited eventually at a genetic level. If new facilities allowed a greater potentiality of specific behaviours, attitudes, and skills, and these had relevance for survival, there is obviously a basis apparent for an acceleration of the evolutionary process.

In the preceding chapter (page 166), I have emphasized the importance of the skilful use of weapons and of the invention of more lethal weapons in favouring survival of individual or group and ascribed a major role in human evolution to these factors. This emphasis on behavioural skills does not of course exclude the development of physical qualities of strength and endurance as being also highly important for survival in combat. Another feature that could well have sprung from a similar source is the concentration of human female evolution on sexual attractiveness in the young at both anatomical and behavioural levels. To be the mate or one of the mates of a specially successful warrior is clearly enough biologically desirable, and that type of evolution must have gone on more or less in parallel with the development of male skills and attitudes with a military bias.

Essentially, any genetic changes in neural infrastructure that could be seized on to modify behaviour in inheritable fashion would be of special importance where behaviour had a particularly direct bearing on the liklihood of producing offspring that would survive. Survival in combat for the male and sexual attractiveness for the female are the obvious examples to be expected in primitive hunter-gatherer society and through most of subsequent history.

When urban civilization developed, the new opportunities for the deployment of latent genetic capacities, discussed on page 179, with probably some associated improvement in survival for offspring, should have ensured a continuing influence of behavioural factors on the course of human evolution. When we reach this point, however, accumulation of mutations can almost be forgotten as a significant evolutionary factor. It may have been a major influence in determining changes such as brain enlargement taking place over a million years, but not within the less than 10 000 years since the first cities arose in the Middle East. For the shorter period recombination of existent genes is the only source available for significant hereditary change involving substantial groups of people. One can hardly even dream of a controllable genetic process by which the effectiveness or diversity of recombination could be modified. However, as in all matters of population genetics, if a mutant gene can once be safely installed in the gene pool and offer selective advantage for survival, it will inevitably change the character of the species or at least the mating group.

So far, we have been concerned only to account for the speed with which inheritable aspects of human faculties have evolved within a mere one to two million years and to find some basis for the incredible diversity of human potentialities. But all through this book I am at least as much concerned with ageing and death as with survival, and with genetic disease as well as a healthy inheritance, and, in my next chapter, with crime and other aspects of evil in so far as they have genetic associations. Men and women are as diverse in their faults and foibles as in their faculties. Mammals and birds in the wild seem to be always at a peak of health, fitness, and beauty, and idiosyncrasy of behaviour is as rare as an unthrifty coat or bedraggled plumage. Whenever large groups of children or adults are closely examined for their fitness for some social requirement—with military service the prototype—an unexpectedly high proportion of physically unfit, mentally dull, or emotionally unstable subjects are found, and the diversity of the ways in which they are found wanting is as great as the diversity of talents among the healthy and acceptable. In the only animal population where selection for survival has been muted almost beneath recognition, such a diversity of harmful or unwanted characteristics is the inevitable result of any type of error-prone process that may be specially characteristic of man.

However, all the discussion in the last few pages in support of an hypothesis that man is subject to error-prone handling of certain aspects of DNA beyond any other mammal must remain unconfirmed and infertile until some molecular genetic evidence of the type of process postulated has been provided. Until then, the possibility must remain that human diversity is basically no more than that in any other animal and becomes evident (a) because of our uniquely protected survival, and (b) our ability to recognize and magnify relatively minor differences within our own species, especially in regard to behaviour.

14

Of Good and Evil

It is no more than a simple syllogism: to be good is to react in the normal fashion laid down by the genes toward one's kin and, at one remove to other members of the in-group; to be evil is to exert against a member of the in-group behaviour that was once appropriate toward the out-group. Our evolutionary approach allows no other biological basis for human ideas of good and evil, but it can only be the beginning for an understanding of the major human problems of personal codes of behaviour, of dishonesty and crime, and of social order and government. In the primitive hunter-gatherer community there was a clear differentiation of *us* from *them*, of the in-group we know as individuals from the out-group of strangers and potential enemies. Even at the primitive level the difference is not one of absolutes. Close kinship, particularly the mother-young child relationship, provides a stronger bond, a more affectionate relationship, than is formed with those less closely related. And circumstances will always be liable to modify the intensity of the dislike for alien individuals or groups.

In attempting a discussion of ethics in the light of developing ideas on the importance of genetic factors on human behaviour, it will be even more necessary than elsewhere in this book to present a positively naive simplification of the complexities of the modern world. I am concerned only to seek ways by which one's working knowledge of biology in general, plus a specialized acquaintance with aspects of human disease and human genetics, can be applied to help understand the main social and ethical problems of the present day. I must repeat that in such writing there is no possibility of being certain and that one is concerned above all in producing something that anyone with university-level intelligence (conventionally an I.Q. of 115 or higher) and reasonable education can understand and evaluate. If the approach has any virtue it is that it is based on disciplines that are in principle subject to progressive clarification by accepted scientific methodologies. Discussion of the human situation from Plato to Bertrand Russell, by philosophers or politicians without understanding of human genetics in the light of molecular biology, ethology, and evolutionary theory, must be totally inadequate, even if this is certain to be denied by every critic educated anthropocentrically in the humanities. Such a claim for the inadequacy of a purely humanistic approach to ethical problems does not of course imply any positive claim that the present discussion has any more scholarly virtue than any other speculative development of a scientific field. It is to offer something to be modified or discarded with every new addition of relevant biological information.

Kinship and altruism

The possibility of altruism arising in the course of evolution has provoked discussion ever since Darwin published *On the Origin of Species*. The social insects provide examples, but they are remote from any human analogies. The beehive or ant nest are essentially superorganisms, and survival, to be of evolutionary importance, must be of the community rather than the individual. A more acceptable approach, initiated by J. B. S. Haldane, depends on the fact that altruism, even to the extent of self-sacrifice within a kinship group, may be the most successful way of ensuring the survival of genes under appropriate circumstances.[1] It must be remembered too that altruistic behaviour, involving or threatening the death of the individual, is probably very rare, and when it does happen will occur as an episode in the course of normal care and protection of the young that must always entail a certain evolutionarily acceptable risk of death.

When agriculture emerged and cities became possible, the most important social change may well have been the great enlargement of the average size of human communities. Under the new circumstances it became impossible for one person to know as individuals more than a small proportion of the people he would meet in the urban environment. So the city-based civilization of the Middle East became the forerunner of modern 'anonymous society', to use Eibl-Eibesfeldt's expression. In a large contemporary community a person's kinsfolk will nearly always represent the core of his in-group, to whom he will be bonded be genetically based behaviour traits which will include the possibility of altruism. Inevitably, the bulk of the population will be strangers and essentially an out-group whom it may be expedient to tolerate. Law and order has had to come into being to ensure that behaviour toward anonymous individuals within the larger community must be more or less equivalent to that proper between kinsfolk. Some effort must always be called for to achieve this and sanctions will need to be applied by any ruling group to ensure it. In such a situation, genetically based bonding toward kin with its implicit evolutionarily justified potentiality of altruism and self-sacrifice has to contend with an equally innate tendency for aggression and violence toward strangers. At the same time the presence of authority and an instinct for survival plus a commonsense understanding of the requirements of a functioning social organism will inevitably produce the cultural compromise with genetic predispositions, out of which systems or personal ethics and public law have been developed. The compromise has always been an unstable one, prone to manipulation by aggressive and intelligent men. The art of politics, particularly in its expression as war, is to be able to twist normal judgements, with their combined genetic and cultural origins, as to who forms part of the in-group and who are of the alien out-group. It is always necessary in power politics to bring people to feel that they can identify themselves with some new political grouping and to intensify their hostile attitude to some out-group. It is standard strategy to foster unity and purpose in the in-group by emphasizing the alien character and aggressive intent of the chosen adversary.

At an early period, probably as soon as it became necessary for intelligent men to handle the organization of city states, these problems of human conflict must have become acute and the general nature of the difficulties must have been recognized. Ways had to be found to deal with them. Kings or military oligarchies sought to control by threat of violence, priesthoods by the application of more subtle methods out of which organized religions and systems of ethics were to develop. History on this view is concerned almost wholly with male aggressiveness and the intelligence that can plan its effective use for the seizure and maintenance of power. Any power that forces people to behave in fashions unnatural to their innate responses is to that extent evil and only to be tolerated if it is exerted to the overall satisfaction of the community and is associated with loyalty to the source of power, individual or institutional, by a substantial proportion of the people. The traditional dictum that all power corrupts, and that successful extension of power whets the appetite for more, is borne out by history and has its biological analogies from laboratory studies of aggression in mice, which I have mentioned already. A hen, too, will continue to fight her way up the peck order until she is defeated.

Bonding and goodness

Conversely, to be good is to widen the in-group, to which normal genetic potentiality allows automatic bonding and sympathy so as to include many others—in the limit of saintliness *all* those outer groups to whom the normal genetically based attitude is one of hostility. Goodness is wholly a social relationship, and its first requirement is a positive interpersonal bond with the individual to whom the goodness is directed. The classic image of goodness is the young woman with her infant at the breast, and it extends naturally to care of children and the extension of maternal solicitude and attachment to the rest of her kin, not excluding her husband-mate and the old folk. Goodness is preeminently something to be looked for in women, for in primitive times the role of woman in favouring group survival depended more on this quality of goodness than on any other. Goodness, capacity to generate positive bonds with any individual identifiable as of in-group, was as important for the woman as aggression was for the man. For a million years at least, *his* most significant role was to defend his group by killing those who would destroy it. Interpersonal bonding is not so easy for a man except where it is reinforced by a sexual relationship. There are of course many men with a natural facility for friendship and it is normal to find a strong parental bond, which however is eventually lost either with dignity and goodwill or catastrophically after puberty in the child. It is probably a valid generalization with only minor exceptions that between men firm bonding is only possible where there is an accepted and fairly stable dominance-subordinance relation between them. It is characteristic for a scholar of stature, a politician in high office, or an army commander to love and take pride in the achievement of a brilliant young man under his aegis in his own profession, but only so long as the dominance relationship remains unbroken. In any normal male, aggression will always overcome bonding when any ambivalence develops in the situation.

My own experience and observation tell me that the most important function of alcohol in company is to diminish the potential aggression and dislike between men who are not securely within each other's in-group. And for the corresponding reason alcohol is much less necessary in female groups.

Goodness today

Any serious discussion of contemporary ethics would have to be deeply concerned with the changed code of sexual behaviour and the disappearance of most moral and legal sanctions against several types of activities previously frowned upon. It would be inappropriate as well as being outside my competence to deal with that topic. My whole discussion must not move too far from the central theme of the nature of ageing and of related influences of genetic error on human function.

As one who has never been seriously in debt, has never been worried about losing or seeking a job, and has been ambitious only within a very restricted field, I have had no difficulty in following a law-abiding path. Much the same probably holds for 80 or 90 per cent of established academics and professional people and for such people goodness has become almost meaningless. It becomes little more than 'behaving nicely to people' according to the current habits of one's class. Probably there is more to it than that, and I find myself attracted to a simple-minded list of how it is appropriate for a civilized man or woman to behave in order to facilitate the smooth running of the community. They were put together by the late Dr L. J. J. Nye of Brisbane from the common wisdom of the ages in the form of six axioms which I quote from him:

(1)   Take what fate ordains.
(2)   Be honest.
(3)   Do the best you can.
(4)   Be kind.
(5)   Be cheerful and enjoy living.
(6)   Practise economy.

I am certain that it is good advice and a fair indication of how most likeable people not under stress do in fact behave. But where there are genetic flaws, poverty, illness, and other disadvantages, those rules may be very hard to follow. One has only read over the list to see that the axioms come very close to embodying the traditional virtues of the working-class wife of fifty years ago. Unfortunately they have a decidedly old-fashioned sound in the 1970s, and I wonder whether the average male could consistently follow them without a strong effort of will, bolstered by some psychological support from outside himself. I am impressed by the suggestion that England escaped a violent revolution by the success of the Methodist revivalism under John Wesley in the eighteenth century, with its care for the disadvantaged and its precepts for personal behaviour.

With the disappearance of any sense of the importance of religion in the vast majority of educated people, there is a lack of any contemporary support for goodness of the old-fashioned variety typified by Nye's axioms or the golden rule. As judged by their writings, the main ethical imperative of articulate young people in Australia and other Western countries is a special care for the disadvantaged, which in practice takes the form of supporting political or violent action by sections of the disadvantaged. The downtrodden must always be right, no matter how antisocial their behaviour or how blatantly unintelligent their attitudes. In a few instances this bonding to those who are black, impoverished, genetically crippled, or old is genuinely felt. From what one can gather from direct experience and from the media, far more of that assumed concern is more closely related to aggression than to love. The commonest source appears to be the hatred of the aggressive toward an individual or group *outside* who have privileges that they lack. To transform that hate into concern and compassion for those whom the hated group or that group's ancestors have by repute maltreated and disadvantaged has some obvious psychological advantages for the hater. In particular, it gives a justification for hating, and to some extent feeling superior to, the privileged group.

Again, it may be the prejudice of the old and the unaggressive that sees 'demonstrations' as wholly primitive reversions to the war dances needed to whip aggression to the killing point—something that may have been necessary for survival once but is utterly inappropriate today. For an aggressive and unintelligent male it is extremely difficult to be good in the sense of being valuable to society and capable of 'loving care' to others, just as it is easy for his aggression to be lit up to violence in company with others of his type. Vandalism springs from the same source, and it is universal experience that when a political demonstration gets out of hand, cars are smashed and burned and shops looted and set ablaze. Until the balance of the genes that gives us human nature is changed, the male *lumpenproletariat* of the nineteenth-century revolutionaries will always be available to pull chestnuts out of the fire for cleverer evil men.

Quite possibly because of some deepseated abnormality in my own temperament (of genetic origin, no doubt), I have found it supremely difficult to think of a 'good' action that I or any other male human being has done that was more than an in-group response of kindness and care primarily, and sometimes exclusively, to kinsfolk, and especially to one's own children and grandchildren, but extending variably outwards. Protestations of love for everyone, and particularly for the poor and afflicted, always impress me as rank hypocrisy, always part of a play for power. At the risk or the certainty of being accused of adopting a ridiculous nineteenth-century attitude worthy of Comte or Herbert Spencer, I believe that goodness in men can only become evident as a result of intelligent action. Instinctive goodness is the prerogative of women, of a proportion only of women.

The possibility of intelligent goodness

My own sense of what is good is largely determined by my personal history. When I had made the grade and felt that I had become a competent research worker in the field of infectious disease in the early 1930s, I can recall having no slightest doubt that what I was doing was unequivocally good. As an agnostic scientist and a Fabian socialist in politics, I had the normal contempt for the Establishment, but I cherished the feeling that I could look anyone on earth in the eye and feel certain he would approve of what I was doing. I have nothing like that certainty today, but between the wars it was completely sincere. I felt that if by devising methods of vaccination against influenza or polio I could save lives or prevent permanent disabilities, I must be doing good. In retrospect I think some of that rather smug enthusiasm really came from the peculiar gratification of successful work in the laboratory.

When a man experiences a sense of clean satisfaction that he has accomplished something that required disciplined effort and the exercise of his physical or mental capacity to the limit, and which harms no one, he cannot avoid feeling that he has done something *good*. It could be to climb Everest, to capture a ten-foot black marlin, to write a sonnet, or to identify the micro-organism responsible for some queer fever. If the action falls into the category of admired achievement and wins prestige or tangible reward, the satisfaction is greater but that is not a necessary element. I have found the same keen sense of elation and 'goodness' when I have found an important new technical paper, followed a difficult argument, and been convinced that here was something really good that in larger or smaller way was going to modify significantly my understanding of that area of science. No credit attached to me, but the excitement that the author had left between the lines had been communicated and appreciated.

Most people will probably object that this particular attitude is quite different from the more conventional form of goodness that I have been discussing. Many would in fact regard it as sheer vanity and self-righteousness—which, in parenthesis, are very difficult words to translate into anything biologically meaningful. Nevertheless, there is, I believe, a real relationship between the elation of achievement and the indefinable but always recognizable feeling associated with having been helpful in someone else's trouble. In scholarly and scientific achievement there is a sense of unity with what is the most important of one's in-groups, the world-wide community of scholars within and on the periphery of one's field. When anyone nowadays talks of service to humanity, he is immediately suspect of being subtly or blatantly dishonest. Judging from my own experience, the sense that one is potentially or actually helping to alleviate pain and death, or to improve the life satisfaction of some vaguely pictured mass of other human individuals, is more often the product of goodwill and a certain naivete rather than a half-cynical self-deception.

A closely related modern basis for good action, with an intelligent rather than an interpersonal form, has developed with the recognition of the need for conservation of resources, the protection of the biosphere from pollution, and so on. It can be made into something like a religious justification for a modern ethic, and at the present time it is an increasingly important factor in political and industrial decision making. I have written and spoken much about the need to develop a stable human ecosystem for the earth, and the implications of that concept are behind every chapter of this book. On one occasion my Presbyterian upbringing emerged in an attempt to answer the first question of the Shorter Catechism—'What is man's chief end?'—in terms acceptable to the twentieth century. The answer is quite different, but in its own way probably means much the same as the Assembly of Divines at Westminster intended in 1647. 'Man's chief end is to cherish his home, the earth, and enjoy its bounty for ever.' In other words, those actions are good which help to ensure that all future generations of men shall have a world that can fulfil their needs to at least the extent that ours have been, and that the health, intelligence, and capacity for goodwill of those descendants will not be inferior on the average to our own. That has no tang of Utopianism, and the criteria for a good life are in principle all objective and most of them measurable.

Any realist will probably object that such an aphorism is only another example of a universally acceptable set of Utopian principles for the future that is characteristically voiced as preamble to some scheme for the short-term enrichment or aggrandisement of a covey of political plotters on the left or the right. Life, liberty, and the pursuit of happiness in America, or the temporary dictatorship of the proletariat that will eventually dissolve into a society so satisfying and tranquil that government will not be needed in the U.S.S.R.—somehow it has not quite worked out like that in either. To concentrate for all time on the maintenance of the biosphere and a concern that the biologically significant qualities of our species should not degenerate has, however, a totally different quality from those. In particular, it provides a principle that can be applied by any humanly conceivable form of social control and it can never become outdated.

In the last chapter of this book I shall give my reasons for believing that it will take not less than a million years to produce a human species that could be genetically adjusted to physical and mental health and a dynamic social stability in a world of today's type in the cities of Australia, America, and Western Europe. And of course no one can predict that human evolution will take the course that an unaggressive biologist, writing in 1976, would like it to follow. I have no doubt that if there are biologists in a million years' time they will have found clear evolutionary reasons for whatever is then the human or posthuman situation, but it cannot be seen from the other end of time. All that can be attempted must be in the light of current knowledge and aimed only at fulfilling felt needs and righting abuses as they become intolerable without in so doing lowering future genetic integrity of the species or significantly harming the environment. In a world where little can be seen or read of that is not motivated by personal or institutional greed for money or power in the short term, that may be impossible. There are no intelligent Utopians nowadays; revolutions have always led to tyranny worse than what was overthrown. If we are convinced, as most of us are, the Popper's 'piecemeal social engineering' is the only reasonable and workable approach to social change, we can have little hope that changes over the long term can be directed intelligently.

Intuitively I feel with Sir Charles Darwin that the next million years will be an intermittently devastating series of bloody and gene-deforming conflicts in an overcrowded world that may sometimes bring the human species to barbarism and near extinction. Even the worst imaginable nuclear war, involving all the militarily significant countries of the world, would be most unlikely to exterminate man completely or to destroy the whole of the technical and scientific information that we have accumulated and recorded. The dominant species after the holocaust will still be of human descent, and, even if it takes 10 000 years to reconstruct the biosphere, some sort of ordered social life will re-emerge. Even brief new golden ages of human achievement may occur.

But no matter what is waiting in the long term, the same imperative will remain: to cherish the biosphere and to establish conditions that will give opportunity for the maintenance of the physical health, intelligence, and capacity for life satisfaction of all succeeding generations of men.

Genetic and environmental factors in crime

Experience everywhere indicates that well over 50 per cent, perhaps nearly 90 per cent, of young males are potential criminals, particularly when in company. Capacity for violence is maximal between puberty and marriage. This holds for political terrorism, for automobile accidents, and for all types of violent crime. The steadying effect of marriage is probably only indirectly sexual in origin. Acceptance of responsibility for wife and child is presumably a much more important factor.

On the basic hypothesis I am pursuing, the strength of the genetic components basic to aggression and violence will vary greatly and be distributed over any male population of ages fifteen to twenty-five in the form of a Gaussian probability curve. The immediate occasion of any episode of violent or non-violent crime is necessarily environmental and will almost always include a complexity of relevant factors. It is the interaction of nature and nurture, inheritance and environment, that is responsible for those serious antisocial actions that we label as crime. Any serious analysis of the problem and search for remedies must take account of both.

At the genetic level the primary need is to recognize those individuals in which genetic deviation from the norm is too extreme to make any sort of social adaptation effective. It has probably become impossible to use legal killing of such people as a protection of society. Separated from all the connotations of punishment and revenge, painless and private execution after a careful analysis of the position would be the logical approach. In a community that had rid itself of irrational attitudes toward death, this could conceivably be accepted, but no such community is possible while the human species retains its existing genetic makeup. At the present time a majority of adults, both in affluent and in developing countries, probably support the sanction of 'capital punishment' for actions that gravely threaten society, particularly brutal murder, killing of bystanders or hostages by terrorists, or military treason. Punishment, except where it is used consciously as a negative reinforcement in Skinner's sense during a process of education or rehabilitation, is a residual from the hunter-gatherer phase. Demand for a man's execution represents simply fear, hostility, and desire to hurt and kill one who has become conspicuously a member of the out-group. It is liable to be specially pressed when the offender has previously been openly or tacitly accepted as a member of an in-group. Actions based on fear and revenge can be rationalized in various ways, and on occasion they represent the only way by which a critical situation can be resolved. The only current alternative is imprisonment for life, either in a prison or a mental hospital.

A standard modern suggestion would be to apply one or more of the various therapies that have been applied to psychotic patients showing violent tendencies. The difficulty here is that therapy is necessarily an interaction between the patient and the therapists who are also men with their own genetic individualities of thought and behaviour. The history of treatment of mental disorder is far from reassuring, and with some outstanding exceptions physicians attracted to psychiatry tend to be less balanced in their interpersonal relationships than the majority of medical graduates working in other fields. One is left with the easy answer that much more research on the problem is necessary, in itself almost a counsel of despair. Many things could be tried: various types of neurosurgery, castration for sexually violent crime, electro-convulsive therapy (ECT), and the use of pharmaceutical agents such as hormones, lithium salts, and major tranquillizers. Because of the practical impossibility of ever applying a scientifically based test of the effectiveness of any of these or other procedures, I am completely pessimistic.

The XYY pattern of chromosomal abnormality

About ten years ago it was observed that chromosomal abnormalities of a special type, XYY, in which there are two male sex chromosomes instead of the normal one, were unduly common amongst prisoners incarcerated for violent crime and mental instability. Most of the XYY individuals were also noted to be large men, over $1 \cdot 8$ metres (6 ft) in height. This finding has been confirmed amongst similar groups of convicted persons in other parts of the world. In addition, it has been found that routine examination of the chromosomes in the cells of newborn male infants (karyotyping) indicates that one in about 4000 has the XYY character and that several adult males of XYY karyotype have been found who are leading normal lives with no suggestion of criminal tendencies. Finally, it is highly relevant to the present discussion that a scientifically and socially valuable project to follow the subsequent medical and social history of a considerable group of XYY infants has had to be abandoned owing to the harassment of the investigator by people acting in the name of human rights. The whole XYY story underlines the extreme difficulty of doing anything effective to handle the problem of human aggression. Several points may be made.

(1)   The possession of two Y chromosomes makes it much more likely that a man will be convicted of and imprisoned for criminal and uncontrollable violence than another with the normal constitution. The genetic abnormality in our terms is one of the factors that can bring people to the high aggression side of the probability curve, but it is clear that for it to be expressed other still unknown genetic factors are required.

(2)   For technical reasons, abnormalities in number and form of chromosomes are the only forms of genetic abnormality that can be observed microscopically. They probably all involve a number of genes in a fashion not greatly dissimilar to many polygenic abnormalities which cannot be detected by current methods of karyotyping. Most aggressive criminals show a normal XY constitution and the genetic background of their behaviour must be sought in some other type of genetic deviation.

(3)   Every one of us carries a small quota of genes which in double dose or in the presence of some anomaly in other genes could produce medical abnormality. The recognition of a genetic abnormality in an effectively normal individual is therefore no justification for subjecting him to any type of social restriction. In view of the inevitable psychological effects of knowing that one has a potential genetic defect, it is equally unjustifiable to make such a finding known to the individual concerned.

(4)   From a simple-minded point of view, we should obtain an extremely useful addition to knowledge if 100 infants detected as XYY soon after birth were followed through life and compared with 100 appropriately chosen normal XY infants. On completion such an investigation could allow geneticists to make a quantitative assessment of the effect of an extra Y chromosome on various aspects, good and bad, of the quality of aggression and no doubt other effects on behaviour would be detected. An intensely interesting monograph would result that could undoubtedly influence thought on genetic components of crime and other forms of aberrant behaviour for decades. Given a proper set of normal 'control' individuals and perfect discretion and integrity amongst the examining doctors and psychologists, the investigation *might* have been carried out without the subjects being aware of their genetic anomalies. The objectors, however, were almost certainly right that in this era of mass media and free communication the nature of the study and the names of the individuals would soon become common

knowledge. No child included could escape an implicit smear that could well distort his life beyond anything necessarily predetermined by his genes.

One can feel certain that any other future discovery of a demonstrable genetic factor concerned with aggression will present similar difficulties. Neither a full understanding nor the effective social control of crime are ever likely to be achieved. In the light of man's evolutionary history, evil is probably as inevitable as genetic disease.

15

Vision of a Million Years

One of the classical English examples of the inheritance of high intellectual distinction is the family of Charles Darwin and Emma Darwin, his wife, who came from another distinguished family, the Wedgwoods, of pottery fame. Darwin's grandson, Sir Charles Darwin, F.R.S., was an eminent physicist who had, as might be expected, an additional interest in the biological nature of man. In 1952 he published *The Next Million Years*,[1] in which he speculated about the long-term future of the human species. He chose that length of time as being about what was required to allow a mammalian species to change to and be replaced by a recognizably distinct species. It is an estimate that requires many qualifications, but for a large primate that has been and is constantly changing its style of living it is probably of the right order of magnitude. A million years ago the commonest and perhaps the only extant type of man was an early *Homo erectus*, from whom both Neanderthal man and modern man descended. Anatomically, the earlier species showed striking differences from *Homo sapiens*; the brain was considerably smaller and one can safely assume that average intelligence and manipulative skill were much inferior to present standards. Any modern zoologist would certainly regard them as two different species—though undoubtedly closely related. The range of blood group antigens in *Homo erectus*, for instance, would almost certainly have been nearly identical with ours and similarly distributed amongst individuals of the species.

If we move forward to Cro-Magnon times 30 000 years ago, we find typical *Homo sapiens* with a brain as large as ours and with skeletal details that suggest that his descendants still persist in southern France. Culturally, however, Cro-Magnon man was a hunter-gatherer, differing from his predecessors over a million years only by a richer mental world of mythology and an interest and skill in depicting in paint or clay the large animals of the chase. Since then, what from the point of view of modern European man can be regarded as three major cultural and ecological revolutions have occurred. Initially each involved only a small fraction of the species, but each has spread to include more and more of those previously unaffected. They were: (1) The development of agriculture and the rise of urban life beginning about 10 000 years ago. (2) The rise of a capitalist economy and the application of technical and managerial skills to production occurring around the eighteenth century. (3) The rise of the modern industrial and welfare state based on the conscious use of science after World War II. Behavioural changes of similar ecological significance in other mammalian species could only have been associated with long-term evolutionary change and structural alterations that would differentiate widely divergent species.

The first fourteen chapters of this book are concerned with the nature of man at genetic and behavioural levels in relation to the human and scientific problems of age and death. In this chapter I try to look at future possibilities as to how the interaction of genetic changes and cultural modifications of behaviour could influence the future evolution of man. I shall assume that genetic engineering, involving manipulation at the molecular level, will never be applicable to man and that, therefore, the time needed for change to a new species—or more than one—will be of the order of one million years. If I am right, that socially significant behaviour and misbehaviour now is to a very large extent the phenotypic expression of genetic patterns laid down over at least two million years of hunter-gatherer life, the chance that they will be replaced by something widely different and, hopefully, better suited to a modern type civilization is most unlikely inside a million years. At the risk of being proved wrong 'before the ink on the paper is dry', I feel moderately certain that by 1975 the outline of science as it relates to human affairs had been accurately sketched and most of the practicable applications for worthwhile human benefit had been available for some time. Even more than when Darwin wrote in the early 1950s, if prediction of this sort is possible at all, we probably have most or all of the background of physical and biological (including genetic) knowledge that will ever be available for such an attempt. As will have been evident in earlier chapters, only a very modest start has been made toward understanding social behaviour in man; nevertheless, the sciences of ethology and animal behaviour have helped a great deal. I believe that with an adequate analysis of needs for a humane and reasonable society, with a healthy balance of conservation and order on the one hand and creative non-violent change on the other, a fairly clear idea of the necessary 'mix' of human qualities might be worked out. To do that, however, is very far from believing that such a 'mix' could be achieved, either by human action or through inevitable evolutionary processes. At any given time in human history, one can only frame one's questions and aspirations in terms of currently available knowledge and of the existing climate of opinion. Within those limitations, our problem can be stated as: (1) to envisage the needed genetic and cultural changes in the human species that would make more likely its indefinite survival within a tolerable and scientifically based civilization; (2) to discuss whether such genetic changes could take place without conscious direction during the next million years; and (3) whether any programme of human action could help to guide or facilitate human (genetic) changes in such a direction.

With that as our basic problem some second-level questions need first to be discussed and answered.

(1)   Why is it necessary to rule out any possibility of genetic engineering?

(2)   (a) Could any present or foreseeable human community tolerate a eugenic approach; and   (b) If it did become tolerable, could a eugenic approach be organized with any chance of success?

The inadmissibility of genetic engineering

The classic example of the simplest type of genetic disease is sickle-cell disease, in which the β-globin chain has a GLU amino acid residue replaced by VAL. This in turn requires only a single nucleotide change from T to A in the DNA of the structural gene; in the relevant portion of the DNA the triplet pair in the normal gene
... CTT ...
... GAA ...   becomes   ... CAT ...
... GTA ...   in the sickle-cell mutant.

On paper it seems to be only a simple matter to remedy the error; just reverse TA in the centre pair so that each is back in the normal position on the proper chain. One does not have to think very long to sense the impossibilities. The wrong DNA is in one of the two structural genes that control the nature of adult haemoglobin, and there is only one such molecule in each of the millions of stem cells and the countless billions of haemoglobin-synthesizing cells that are being produced every day. There is no possibility that the change could be made by any sort of chemical approach. The only means that has been so far suggested is that a virus should be 'trained' to take up the right piece of DNA from normal cells. Very large numbers of these carrier viruses would then be injected into the affected patient with the hope that the DNA would be incorporated correctly into at least a high proportion of the stem cells from which haemoglobin-producing cells arise. Some experiments are on record in which it can be shown that cell cultures from a patient with another type of genetic disease could, after treatment in this fashion, produce a few apparently normal cells. This plus much more extensive experiments with bacteria and bacterial viruses is the main justification for those who think of genetic engineering as likely one day to be as acceptable a part of medicine as kidney transplants or open-heart surgery is today.

In all the positive experiments with cell cultures or bacteria, large amounts of virus, a proportion of which were carrying the right gene, were added to cell or bacterial cultures, so that every cell could be infected. Then appropriate tests were made to see whether any of the infected cells had been converted to the desired type. In practice, highly selective methods must be used to allow a minute proportion (1 in 10 000 to 1 in 1 000 000) of transformed units to be detected amongst the mass of unchanged cells.

The odds against success in applying the equivalent method to a patient with sickle-cell anaemia are overwhelming. The virus to be used would have to have a 'cancer virus' quality to allow it to carry the 'good gene' into each stem cell, and it is hard to believe that any official drug authority or other controlling body would ever license the use of such material. Even if the virus particles had taken up each a few genes from normal cells, in not more than 1 in 10 000 would they have received the right gene. Virus particles injected would certainly be capable of infecting many other cells than the stem cells of the red cell series; most would be removed by phagocytic cells, and, even with a large dose, not more than 1 per cent of such cells would be infected, probably very much less. It must be underlined that mammalian genes passively carried by a 'cancer virus' do not multiply when the virus multiplies. The chance of a virus unit with the right gene entering the right type of stem cell and the gene it carries displacing the 'bad gene' and being inserted in its place could be only minimal. Even if a few cells were so changed, the tens of thousands unchanged would continue to dominate the quality of the haemoglobin being produced.

The only approach that has been effective in man applies to a condition in which the genetic defect is the absence of an enzyme needed by cells in many parts of the body but produced solely in the kidney. Removal of one kidney and transplantation of a normal kidney from a suitable donor has in at least one instance been successful for such a condition. But it can hardly be styled genetic engineering.

Eugenics in the long term

In Chapter 8, in discussing the concept of eugenics, I had to admit that in the present state of public opinion it was futile to advocate any deliberate attempt to 'improve' the quality of a human population. Positive action on human genetic difficulties must be based solely on humanitarian care for the individual. Genetic counselling before marriage, and advice on amniocentesis and abortion, are fully acceptable when individuals are known to be at risk, but no such actions would have any significant influence on the quality of human gene pool. In the long term, however, many things could happen that might demand a much more positive attitude to human genetic quality. In looking forward for a million years, we can in principle liberate ourselves from contemporary attitudes and taboos and perhaps come to see some virtues in Francis Galton's ideas. What he advocated was essentially the elitist concept, that people of his own type, healthy, intelligent, law-abiding men and women of goodwill, should so arrange things that their children would marry similarly desirable individuals and, by having families larger than the average, gradually improve the standard of the race.

In the 1970s the idea of eugenics is completely out of fashion. An elitist concept advanced by a mid-Victorian upper-class gentleman can have no place in the political philosophy of modern democracies. Hitler's racist policies are still anathema, and any open support of a eugenic policy would be regarded as equally objectionable by all who write for popular consumption. Just under the surface, however, I believe there is still a lot of sympathy for the Galtonian approach. Parents are gratified when a pretty and intelligent daughter brings home a young man whose innate virtues make him a fit match for her. On the whole, the socially successful have first choice for suitable spouses, and, by the functioning of the semi-closed mating groups discussed in Chapter 13, one can be reasonably certain from simple observation that soft-edged unacknowledged castes do exist and are perpetuated, though often with important changes occurring from generation to generation. No one, however, has claimed that the continuing persistence of minorities of people with special abilities and a significant degree of reproductive restriction to their own and similar social groups could have more than a minimal effect on the genetic structure of the population as a whole.

The development of these semi-closed mating groups may be producing in a proportion of Western communities a series of more or less overlapping subpopulations from whom a high proportion of the individuals in the community with genetic potentiality for high intelligence and outstanding skills will be drawn. Undoubtedly many such people will come from outside these recognizable mating groups, but if it is found to be correct that a large genetic element is expressed in such qualities, they will merely accentuate the genetic quality of the group. On our hypothesis, we may well find eventually that a genetically definable subpopulation contains the great majority of those who are needed for the running of a scientific industrial state and for any cultural and artistic distinction it may develop. There still remain, however, 80 per cent of people whose mating groups are not related to any special skills. Here we have to face the question of human futures in their full bleakness. Is this 80 per cent fated to become a larger and larger fraction of the whole, and its effectiveness, as judged either by its potential contribution to the work of the community or its level of intelligence, physical health, and mental balance, to diminish slowly but progressively? On standard biological reasoning, most of those genetic combinations which in unsophisticated groups would have died before reproduction but thanks to all the altered circumstances now survive, go to swell the unselected central pool. What is going to happen there?

As far as my reading has gone, there are no exceptions to the findings of those studying the I.Q. of school children that there is a negative correlation of I.Q. with size of the family. There are many ways or arguing that this does not mean that the less intelligent have more children on the average than the more intelligent and that intelligence is to a large extent an inheritable quality. Still, that does represent the simple commonsense explanation, and I believe that it is the true one. It would follow, therefore, that the trend of intelligence *insofar as it is genetically determined* is downward. This is not inconsistent with the findings that, as far as comparison is possible, there has been a small rise in the average I.Q. of London school children over the last forty or fifty years, presumably due to improved nutrition, better teaching in school, and an 'enriched' environment. On standard evolutionary reasoning, one may safely assume that, during primate evolution up to the end of the hunter-gatherer period, approximately 80 per cent of progeny died before reaching reproductive age. Now, less than 5 per cent of children in Western communities are lost, and this sudden relaxation of the winnowing function of natural selection must result in an accumulation of individuals whom we would call inferior in terms of our current values on health, intelligence, and aggressiveness.

We must remember, however, that in the eyes of evolution there is only one criterion of success: an increase in the number of individuals reproductively active as one generation succeeds another. If healthy, vigorous, intelligent people average two surviving children per family while others of low intelligence and slovenly habits rear four or six to adult life, the latter is the biologically more successful group in the existent environment. Even if its existence is only made possible by payments from the state and supervision by welfare officers of all sorts, the reproductive success of the 'inferior' group marks them as biologically superior. Insofar as the relevant characters are genetic, they would be the sort of people to inherit the future, if the current pattern of the welfare state persists indefinitely.

The validity of the reasoning I have been using is strongly disputed by many authors, particularly those in the fields of education and sociology but including some biologists of high academic status, such as Lewontin. They have the moral and political support of the great majority of those who regard themselves as underprivileged, and there is a major opinion trend that, irrespective of which side is 'right' in regard to the facts, current democratic values and political expediency demand that no differences in genetically based abilities shall justify any differential social treatment of individuals. However unsatisfying such an attitude may be to those who feel that they have genetically based capabilities that the community should foster, they may have no alternative but to accept it for the foreseeable future.

I am content if we can ask the community for no more than the right to go on studying the distribution of human abilities and other socially significant qualities for a long enough time to establish what objective changes are taking place and to evaluate the genetic component in the longer term changes. In all probability, significant genetic changes will first become unmistakable when medically important genetic defects become unpleasantly evident, and, no doubt because of my special professional interests, I feel that a diminution in the average resistance to infection may be the first one to be established.

Population genetics of infection and immunity

During the past year I have been deeply interested in the possible genetic effects of infectious disease as controlled (a) by the genetically determined efficiency of the immune system, (b) by the degree of hygiene, which can be roughly equated with the standard of living in the community, and (c) by medical developments in the prevention and cure of infectious disease.

Before about 1800, population increase in Europe had been slow, with only a minor degree of acceleration becoming visible in the preceding century. With approximately ten or twelve births to the average completed family, only a fraction over two survived to contribute to the following generation. Most of the deaths were in infancy and childhood; nearly all were the result of infection, accentuated by poor care and nutrition and determined primarily by the crowding, poverty and filth that made cities hotbeds of infectious disease of all sorts.

Healthy human existence in cities became possible only with the development of a public health conscience and of practical methods of sanitary engineering in mid-Victorian England. Until then, the cities could never have maintained their numbers if it had not been for the perpetual attraction they held for young people from the healthier countryside. One of the most illuminating of historical sidelights on the factors concerned with human populations comes from a fascinating study[2] of a genetic disease almost confined to South Africa, a form of porphyria whose commonest manifestation is a characteristic type of skin inflammation particularly on exposed areas of the skin. This abnormality had been well known for more than a century to be characteristic of certain Afrikaner families; 'van Rooyen hands' it was often called. The abnormality, however, was associated in some individuals with important general symptoms, and Geoffrey Dean carried out a comprehensive epidemiological and genealogical study of the disease, which is a dominant condition easily recognized and a mark of distinction rather than a stigma to those who carry it. Dean was able to trace the inheritance of the condition back to a single couple, Gerrit Janz and his wife Ariaantje. Janz was one of the first free burghers at the Cape Town station of the Dutch East India Company and his wife was one of the eight orphan girls sent out from Holland to become wives of the young men (free burghers) wishing to take up land for farming in 1688. This was the beginning of the Dutch colonization of South Africa and of a flourishing and fertile community.

In 1950 there were estimated to be about 7000 cases of porphyria in South Africa, almost wholly in people of Dutch stock. Evidence from the distribution of surnames in the country, plus the results of the porphyria investigations, gave a firm indication that a fertile couple starting a farm in 1688 had in less than three hundred years an average of 12 500 descendants. If that couple had lived in the Netherlands, the average count of twentieth-century descendants would not be more than twelve. In the seventeenth century in South Africa, there were no towns and only a very sparse indigenous population when the expansion into the veldt took place. Farms were isolated, food easily obtained; there was little opportunity for infection, and the ten or twelve children grew up healthy to marry and take up new land further out. The same outburst of population followed the eighteenth-and nineteenth-century colonizations of the other empty lands of North America and Australia, but neither produced so clear a 'marker' for the process as the South African porphyria.

In a curious way this story of the speed with which the empty lands were populated by European colonists in the last two hundred years has a relevance to the future. There are no more empty lands attractive to farmers, and the world's worst problem is overpopulation. The only way that situation could be changed would be as a result of the aftermath of a major nuclear war, something that all the logic of history tells us could occur within a century. Even if such a catastrophe reduced the world population eventually to only 0·1 per cent of its prewar level, the present population could be regained in less than two centuries if each completed family averaged eight, which was about the level of the Boer pioneer families.

Any complete, or nearly complete, breakdown of civilization with an enormous reduction in human population would tragically terminate the course of human history for anyone surviving to look at the events in terms of a human lifespan. At an evolutionary level it would be only an episode, with little or no effect on the human or post-human situation a million years hence. Always provided that a few libraries survived and a reasonable number of those human genetic combinations remained that are responsible for intelligence, leadership, and manipulative ability, a world community broadly similar to our own might be functioning within a hundred years. It would differ, however, one would hope, in having developed a determination, as deep and permanent as the instinct for self-preservation, never again to touch nuclear technology of any sort. To me it is axiomatic that no continuation of the human species in a form that any intelligent man or woman of the last 2500 years could contemplate is possible unless we can banish nuclear power and nuclear weapons from the earth.

Eventually, world opinion must accept the two imperatives: that nuclear technology of all sorts must be permanently banned, and that the biosphere, fuelled only by solar energy, shall be maintained in all those respects that are necessary for healthy life. The second imperative implicity includes the first.

Any discussion of the next million years must make the assumption that these imperatives prevail. But it may need experience of the horrors before the decisions will be taken.

HLA groups

To return to the theme of the genetic changes that are now taking place, and possible ways by which they may be recognized. As an immunologist greatly interested in autoimmune disease, I have been deeply impressed during the last five years by the discovery of a relationship between the HLA 'tissue groups' and the occurrence of autoimmune disease in man.[3] An outline of these tissue groups and an indication of their significance has already been given in discussing immunity and ageing in Chapter 6. Omitting certain recent complexities, we can say that each individual has four antigens which are controlled by two sets of genes, A and B, on each of a pair of corresponding chromosomes. An A, B set on a single chromosome, A1, B8 for example, is known as a haplotype. Since all the antigens are expressed by codominant genes, both genes and antigens are called by the same names. Many thousands of people have been tested for the types of these antigens that they carry, in the expectation that the results could have some significant bearing at the medical level. The first really striking result was the discovery that 95 per cent of patients with a rheumatic disease of the spine, ankylosing spondylitis, had B27 as one of their four antigens. In normal people of European origin, only about 4 per cent have that group. The group is therefore quite exceptionally prevalent in persons with this particular disease, and since the four groups we possess are transmitted from parents to children in simple Mendelian fashion, this must mean that the susceptibility to ankylosing spondylitis is also, in part at least, genetic in origin.

In similar fashion it is found that A8 is unduly frequent amongst patients with any one of seven autoimmune diseases, while A7 is too frequent in persons with multiple sclerosis. And we must not forget that at least three other autoimmune conditions have an excess of B27 in addition to ankylosing spondylitis. All these conditions are known, or strongly suspected, to be autoimmune in character. Now the simplest definition of an utoimmune disease is a positive and active response against antigens which in mormal individuals are spared any such attack. It is enlightening that there are some people showing an excess of A8 who suffer an illness not very unlike an autoimmune disease, which results from an immune attack against a foreign protein, but one for which it is essential to health that it should not become involved in immune reaction. This is coeliac disease, an immune sensitization of the bowel against gluten, the main protein in flour. In an oversimplified but, I believe, legitimate fashion, it is convenient to regard these unnecessary reactions against self-components of the body or universal normally harmless materials from the environment as reactions that are *too strong*. This opens up the possibility that if similar strong reactions were directed usefully, against dangerous micro-organisms for instance, they could be advantageous to survival in heavily infected environments.

Antigen B8, then, is present in excess in quite a wide range of autoimmune diseases and in the gluten sensitization we call coeliac disease. The B8 antigen has another peculiarity: it is one of the more commonly found antigens and has a special predilection to be found with A1, another common type. The 1, 8 pair is much the commonest haplotype combination found in Europeans, and the abnormally high frequency with which 1 and 8 are found together is an example in genetic terms of 'linkage disequilibrium'. It is a genetic abnormality for which some evolutionary reason must be sought. I have suggested that the presence of 8, and particularly if 8 and 1 are on the same chromosome, is in some way linked with capacity to make exceptionally active immune responses, responses that are so 'good' that in a small proportion of people they provoke pathological effects like autoimmune disease. If at the same time the 1, 8 haplotype is unusually effective in helping to diminish the mortality from commonly lethal infectious disease, this would provide a reason for its increase beyond a random level. It could be, therefore, that 1000 years of evolution in European towns and cities, when infection during infancy and childhood was by far the most frequent cause of mortality, could have led both to the 1, 8 disequilibrium and to the association of several autoimmune diseases with antigen 8.

If such an interpretation is correct, we should expect that, with the virtual removal of infectious disease as a cause of death in childhood, a slow swing back of 1, 8 to the equilibrium level should occur and the average genetic resistance to infection fall. From the point of view of assessing what genetic changes were taking place, a change in the proportion of people with 1 and 8 toward the equilibrium ratio would be much more accurately measurable than changes in the incidence and severity of childhood infections, simply because of the complexity of the factors governing the latter. When general community living standards, degree to which children are immunized, and competence of medical services may all be very deeply concerned in the result, it is very hard to sort out any genetic factors that may be involved. Relatively gross deficiencies in immunity that are of genetic origin undoubtedly occur and during the last twenty years much has been written about these diseases. It is quite unlikely, however, that this is related to the process I am speculating about. Scientific interest and the possibility of treating cases of immune deficiency successfully have together been responsible for the increasing number of reports. Any evolutionary change associated with our new capacity to prevent death from infectious diseases will be a slow and inconspicuous process. Inevitably some increase in minor immune deficiencies will occur, but modern antibiotic therapy should remain adequate to handle it. One cannot exclude the possibility that in the long term a progressive decrease in the effectiveness of bodily defences against infection could become socially important. For the immediate future, however, the main interest of any change in the 1, 8 linkage disequilibrium will be as a guide to the sort of evolutionary change that may be produced in a modern technological civilization.

Colour blindness

Red-green colour blindness is a well-defined genetic deficiency transmitted by a gene on the X chromosome and therefore almost wholly confined to males. If a normal male marries a phenotypically normal female who however is genotypically a carrier of the gene for colour blindness, their offspring will be normal except for half their sons, who will be colour blind. Half their daughters will be genotypically normal, but half will be carriers of the gene. If a colour blind woman marries a normal man, the offspring will be sons who are all colour blind, daughters who are all carriers of the gene but phenotypically normal. When both parents are colour blind, all the children, male and female, will be colour blind too.

Colour blindness is found in all races, but it is less prevalent in hunter-gatherer groups (2 per cent ±) than in European populations (3 to 5 per cent). Obviously, normal colour vision is important in differentiating edible objects such as fruit from background, or ripe from unripe berries, etc., and in a hunter-gatherer economy normal vision would be of survival advantage relative to the colour blind. In a civilized society, as Francis Galton noted, colour blind people are unaware of their peculiarity until they are specifically tested, and find no difficulty in the ordinary business of life. As a Quaker, Galton was specially interested in the high proportion of Quakers who were colour blind—5·9 per cent as against 3·5 per cent, according to contemporary measurements. He comments that, amongst Quakers, to dress in drab clothing was as nearly universal as to look on the fine arts as worldly snares. 'Consequently few of the original stock of Quakers are likely to have had the temperament that is associated with a love of color . . . most reasonable to believe that a larger proportion of color blind men would have been found among them than among the rest of the population.'

Darlington rates this observation of Galton's as the first step in demonstrating the importance of the mating group concept in the genetics of civilization. In his final paragraph on the topic, Galton makes the point that those who drift out of the Quaker mating group are likely to be those who are not colour blind. 'Dalton, who first discovered its existence as a personal peculiarity of his own, was a Quaker to his death. Young, the discoverer of the undulatory theory of light and who wrote specially on colors, was a Quaker by birth, but he married outside the body and so ceased to belong to it.'

Colour blindness is a condition of very special importance to the central theme of this book in being a well-defined genetic condition inherited in Mendelian fashion, but affecting the individual only in regard to his behaviour—putting the hunter-gatherer at a disadvantage in gaining his food or preventing those aspiring to be locomotive drivers or the like from obtaining the employment they want. I have little doubt that many of the other genetic bases to behaviour that have been postulated in previous chapters will eventually be related in similar fashion to a *demonstrable* genetic anomaly. Perhaps it is worth noting here that two of the best-known genetic abnormalities can be recognized in an unsophisticated community only by a gross abnormality of behaviour, although their basis is now well understood. They are Down's syndrome (mongolism), associated with an abnormal extra chromosome 21, and phenylketonuria (PKU), where the gross mental retardation is associated with the altered metabolism due to a faulty structural gene concerned with the handling of the amino acid phenylalanine in the body. Neither is by any means as directly related to behaviour as the neural infrastructure one must postulate to account for the inheritance of specific qualities of behaviour, but they do add highly significant evidence of how all sorts of genetic qualities can influence behaviour. More important from our present point of view, they suggest that, as we study human behaviour in genetic laboratories, more and more *measurable* qualities will emerge to help us understand what is happening in the human gene pool now and give us some basis for predicting, and perhaps influencing, behavioural and other inherited characteristics that are important for the long-term future of our species. It is only in this very indirect fashion that we can approach the problems of human genetic quality that are bound to plague human civilizations as long as they exist. At any point in history, however, the possibilities of effective action are very limited; they are more restricted by the contemporary sense of human values than almost any other set of social problems. For the last quarter of this century, systematic accumulation of the facts of human inheritance and their scientific and emotionally unbiased analysis are the immediate requirements. Arising out of this there should soon be developed ways by which genetic change can be monitored over long sequences of decades or even centuries. As knowledge builds up, we may eventually find the courage and the opportunity to take tentative eugenic steps to check the progress of at least some types of genetic deterioration.

The long-term future

In a final section it seemed best to come back to Sir Charles Darwin's book and to try to show where my approach differs from his. I believe that his most important generalization in that book is that man can never be regarded as a domesticated animal. Even in the wildest Utopian dream of people specially bred for desirable qualities, there is always a group of people who decide what is desirable for others—always with the reservation *not necessarily for ourselves*. When we study special pure-line strains of mice, we are always being called on to compare their various qualities with those of the 'wild type'; in human populations there are no pure strains, all of us are still wild type, and, as I have repeatedly emphasized, the basis of our behaviour remains suited to the hunter-gatherer phase of long ago. Expressed rather differently, human nature is, for Darwin and the rest of us, the obstacle to any change in social life in the direction that modern scientists, physicists or biologists, would regard as good.

A brief personal recollection may be relevant here. A year or two after the publication of *The Next Million Years*, Sir Charles Darwin was in Australia, and one evening I was his official host when he spoke to a small dining club of scientists in Melbourne. He spoke about his ideas of the future—at that time not nearly so common a topic as nowadays—and we had some good informal discussion afterwards. It was an excellent evening that everyone appreciated. As we broke up, Darwin congratulated me on the wine, and then said something that I shall never forget. 'Burnet, do you think that anything *good* will ever come out of it all?' This book is, I fancy, mainly an attempt to look at the implications of that question without any real hope of finding a positive answer. The processes of evolution, including human evolution, are not to be interpreted in terms of human values.

The next point made by Sir Charles in his book was the overwhelming role of population levels in relation to food supply. Population must always expand to absorb the sustenance available and at the margin starvation controls population. At rare intervals, 'golden ages' appear, when for one reason or another—e.g. a sequence of exceptionally good seasons or a major technical achievement in the transport and storage of meat or milk—food is relatively abundant. We are now just leaving behind one of those golden ages, when the colonization of the empty temperate lands and the steamship provided food for the cities of the northern hemisphere. Each golden age provokes an increase in population to correspond, and the margin of misery is re-established. Darwin's point is irrefutable, but today more emphasis is being laid on sources of energy than sources of food. Darwin did refer to the pending exhaustion of fossil fuels and the eventual need to use solar energy as the primal source, but, thinking of food supply, put much more weight on the relative scarcity of phosphate rock as a source of fertilizer. As a broad picture of the world of the future, he took the 3000 years of Chinese civilization as prototype and model. It has been through all its history a real civilization, with an art and literature almost always far ahead of its contemporary Western equivalents, nearly always with a margin of people at the edge of starvation and periodically subject to flood, famine, invasion, and civil war, but never quite losing its essential quality.

Darwin implicitly took the view that 'human nature' will never change significantly, and, like Popper and other conservative and responsible philosophers, saw no possibility of a Utopian solution. I am sure that he would agree with Popper that Utopianism is both impossible and not to be desired or worked for; social amelioration can come only by piecemeal social engineering.

For me, the only weakness in Darwin's approach is his implicit assumption that even a million years is not long enough for genetic changes that could modify the neural bases of human behaviour to take place. Most of those who are competent in human palaeontology feel that during the last million years man changed greatly in the bony form of his skull, doubled the weight of his brain, and developed both the anatomical modification in mouth and larynx and the neural infrastructure that made language possible. It would be foolish to deny the possibility of quite different but equally significant genetic change in the next million years. I hold too that human intelligence, based on a continuing use of the methods of scientific logic and experimental investigation applied to human genetics, *could* see how to provide the best opportunity, as it were, for spontaneous evolution to move in directions favourable to the fuller expression of the specifically human qualities of our species.

So with what may seem to most people either wholly unjustified intellectual or scientific arrogance or simple-minded Utopianism, I want to finish with a statement of such long-term aims, and what could help to attain them. Clearly, those aims must be stated in regard to both man as a species and the global biosphere which must support him. In principle, a long-term vista could cover up to the four or five thousand million years before the sun fails to produce the precise flux of energy on which life depends. Over that nearly infinite period man's long-term descendants *could* remain the dominant life form on the earth. It is reasonable, however, to think of the million years we started with as the farthest limit for specu-lation, and in fact be more concerned with the next century.

First, then, in regard to man. I should start, as I did in the 1966 Boyer Lectures of the Australian Broadcasting Commission, with four rules for a modern ethic.

(1)   To ensure for every individual human being the fullest measure of health that is allowed by his inheritance.

(2)   To provide for everyone the opportunity to develop intel-lectual and manipulative skills within the limits of his inherited capacity to learn them.

(3)   To ensure for all the opportunity for achievement and the recognition of success.

(4)   To ensure that opportunity to attain mental and bodily health, and to find satisfaction in achievement, will be available to all future generations in a measure not less than what we now enjoy.

Everything in that list has genetic implications. On the side of the biosphere and the other global resources, my 1966 summary still seems relevant.

The resources of the earth must be maintained for the use and enjoyment of future generations by ensuring four things:

(1)   That unrenewable mineral resources shall not be exhausted before effective substitutes can be made from always available materials.

(2)   That energy from fossil fuels shall be replaced in time by perpetually renewable sources of energy.

(3)   That the environment shall not be poisoned by industrial and military wastes.

(4)   That adequate areas shall be preserved to allow the indefinite persistence of all significant forms of wildlife and many areas of natural beauty or special interest.

Care of the biosphere is essential in any regard for the future, but in this concluding summary our primary concern is with human genetics and the changes in the gene pool that we may expect irrespective of whether we can or cannot influence those changes. In the 1966 formulation I tried to provide criteria of what should be desirable in man that were wholly free from those matters of race, tradition, religion, and power, about which otherwise intelligent people can quarrel. But I made virtually no suggestion as to how such aspects of 'human nature' could be modified. Unless they can be, we are left with Sir Charles Darwin's prediction of the future. The crucial requirement for human action would be to find ways to favour selectively higher reproduction of subpopulations in which undesirable excess of male aggressiveness was less likely to appear. I can see no indication of how this might be accomplished at the present time, but I think I can see faintly how new concepts might be developed which in the distant future could have practical implications in such directions.

The need is simply for greater information on the distribution of human faculties and their components of origin, genetic or environmental. If the central theme is correct, that human behaviour and human idiosyncrasy differ in all populations in such a fashion that about 80 per cent of the variance is of genetic origin, the first requirement is to establish that fact and its significance in the mind of every literate individual. Far less than 1 per cent are so convinced at present, and the conclusion may not even be correct. Research, discussion, and public education in human biology are the primary requirements. Most of the techniques are available. What is required is the will and the organizing ability to carry out studies on large numbers of persons plus accessible data on the socially significant history of each individual in the community. The objections that such studies violate the individual's right to privacy and diminish personal freedom will have to be overcome in one way or another. In the past, and probably in the future, it may well be expedient to use the medical implications of genetics to build up public acceptance of the necessary investigations. If research into the factors concerned in mongolism requires the chromosomes of very large numbers of children to be examined for 21-trisomy, there will be incidental opportunity to gain hints as to whether any other chromosomal characteristic has medical or social significance. An approach of this sort has the double advantage of by-passing the need for probing into matters like temperament, criminal record, or sexual history, which are almost automatically resented as an illegitimate invasion of privacy, and of providing unequivocal 'hard' genetic data in contrast to all the possibilities of distortion and lying inherent in obtaining personal information from large numbers of subjects. The need will always be to find some unequivocal marker that is subject to laboratory assay and that has a direct or indirect association with the socially or medically significant behaviour or functional abnormality that we are concerned with. For as long as can be foreseen, continued study of human genetics will remain necessary, with the implied requirement to go on improving the efficiency of the technical methods and the acceptability of such work to the increasing numbers of adults and children who will be needed as subjects. It is always something of a confession of weakness to reach the conclusion that more research is necessary and that until we have much more information no attempt to deal practically with the problem under consideration is justified. It remains true, however, that at any point in a research programme being enthusiastically pursued, new enlightenments previously inconceivable in the literal sense may open up practical possibilities and bring a surge of optimism into the field.

The future of man is the greatest of all the topics open to the scholar, but logic in the working of chance and necessity in biology can only be seen retrospectively. The vast majority of the vertebrate species that have ever existed are extinct, and modern work on the evolution of protein structure, with its emphasis on the importance of genetic drift, strongly suggests that almost every significant genetic change comes from a line that has been at some time on the very verge of extinction. Perhaps the most important weakness running through the whole discussion is the implicit conviction that the dominant position of man at the present time is sufficient to ensure that he can only be removed from that dominance by one of his own descendant species. It is the only possible basis for speculation, but we can never quite rule out the possibility of the extinction of our species without descendants. Accepting, as we must, an indefinite continuance under perpetual change of the human gene pool, our long-term objective as human biologists should be to recognize the direction in which the changes that will determine the form and quality of our descendant species are pointing. Whether it will ever become expedient to take action to foster the changes or to hinder them will depend on too many factors over too many thousands of years to be answerable now.

Whatever will be, will be, and for my part I can be happy enough in the hope that in a million years' time some scholar, still recognizably human, will be talking or writing in broad generalities about what happened to human genes and human health in the first million years of man as a civilized being.

Sources

Having regard to the level at which this book has been written, it would seem to be mere pretentiousness to try to provide full references for every significant statement that I have made. For anyone professionally interested in gerontology, full references to most points can be found in *Intrinsic Mutagenesis*. Throughout the book the behaviour of DNA is central, and my discussions have been largely guided by J. D. Watson's *Molecular Biology of the Gene* (2nd edition). Wherever recent important work is referred to, a reference is given in the list below, arranged by chapters. In addition, I have included a number of general works or significant papers that have had some special influence on my approach.

Introduction
[1] Burnet, F. M. (1974). *Intrinsic Mutagenesis*. Medical and Technical Publishing Co., Lancaster, England.

1   Origins of Life and Death
[1] Watson, J. D. (1970). *Molecular Biology of the Gene*, 2nd ed. Benjamin, Menlo Park, California.

2   Complexity and Error
[1] Monod, J. (1972). *Chance and Necessity* (English translation of *Le Hasard et la Necessité*). Collins, London.
[2] Witkin, E. M. (1974). Chapter VI. Relationships between repair, mutagenesis and survival—Introduction. *Squaw Valley Symp. on DNA Repair*, December 1974. (Preprint.)
[3] Huxley, J. (1942). *Evolution: The Modern Synthesis*. Allen and Unwin, London.

3   A Personal Approach
[1] Robbins, J. H., Kraemer, K. H., Lutzner, M. A., Festoff, B. W., and Coon, H. G. (1974). Xeroderma pigmentosum: an inherited disease with sun sensitivity, multiple cutaneous neoplasms, and abnormal DNA repair. *Ann. intern. Med. 80*: 221-48.
[2] Drake, J. W., and Allen, E. F. (1968). Antimutagenic DNA polymerases of bacteriophage T4. *Cold Spring Harbour Symp. quant. Biol. 33*: 339-44.
[3] Hart, R. W., and Setlow, R. B. (1974). Correlation between deoxyribonucleic acid excision-repair and life-span in a number of mammalian species. *Proc. Nat. Acad. Sci. USA 71*: 2169-73.

4   Ageing in Mammals
[1] Davies, D. (1975). *The Centenarians of the Andes*. Barrie and Jenkins, London.

224

[2] Sacher, G. A. (1959). Relation of lifespan to brain weight and body weight in mammals. *Ciba Foundation Colloquia on Ageing, Vol. 5: The Lifespan of Animals,* eds G. E. W. Wolstenholme and M. O'Connor, pp. 115-41. Churchill, London.

5   Genetics of Ageing
[1] Fialkow, P. J. (1972). Use of genetic markers to study cellular origin and development of tumours in human females. *Adv. Cancer Res. 15:* 191-226.
[2] Orgel, L. E. (1963). The maintenance of the accuracy of protein synthesis and its relevance to ageing. *Proc. Nat. Acad. Sci. USA 49:* 517-21.
[3] Holliday, R., and Tarrant, G. M. (1972). Altered enzymes in ageing human fibroblasts. *Nature 238:* 26-30.
[4] Linn, S., Kairis, M., and Holliday, R. (1976). Decreased fidelity of DNA polymerase in ageing human fibroblasts. (Pers. Comm.)

6   Immunity and Ageing
[1] Walford, R. L. (1969). *The Immunologic Theory of Aging.* Williams and Wilkins, Baltimore; Munksgaard, Copenhagen.
[2] Burnet, F. M. (1976). *Immunology, Ageing, and Cancer.* Freeman and Co., San Francisco.

7   Death
[1] Illich, I. D. (1975). *Medical Nemesis: The Expropriation of Health.* Calder and Boyars, London.
[2] Milton, G. W. (1973). Self-willed death or the bone-pointing syndrome. *Lancet 1:* 1435-6.

8   The Burden of the Genes
[1] Kallmann, F. J., and Reisner, D. (1943). Twin studies on significance of genetic factors in tuberculosis. *Amer. Rev. Tuberc. 47:* 549-74.
[2] McKusick, V. A. (1971). *Mendelian Inheritance in Man,* 3rd ed. Johns Hopkins Press, Baltimore.

9   Xeroderma Pigmentosum
[1] Robbins, J. H., Kraemer, K. H., Lutzner, M. A., Festoff, B. W., and Coon, H. G. (1974). Xeroderma Pigmentosum: an inherited disease with sun sensitivity, multiple cutaneous neoplasms, and abnormal DNA repair. *Ann. intern. Med. 80:* 221-48.
[2] Cleaver, J. E., and Bootsma, D. (1975). Xeroderma pigmentosum: biochemical and genetic characteristics. *Annu. Rev. Genetics 9:* 19-38.

10   Age-Associated Disease
[1] Burnet, F. M. (1957). Cancer—a biological approach. *Brit. med. J. 1:* 779-86.
[2] Priester, W. A., and Mantel, N. (1971). Occurrence of tumours in domestic animals. *J. Nat. Cancer Inst. 47:* 1333-44.
[3] Gardner, M. B., Henderson, B. E., Estes, J. D., Menck, H., Parker, J. C., and Huebner, R. J. (1973). Unusually high incidence of spontaneous lymphomas in wild house mice. *J. Nat. Cancer Inst. 50:* 1571-9.
[4] Peto, R., Roe, F. J. C., Lee, P. N., Levy, L., and Clack, J. (1975). Cancer and ageing in mice and men. *Brit. J. Cancer 32:* 411-26.
[5] Knudson, A. G., Strong, L. C., and Anderson, D. E. (1973). Heredity and cancer in man. *Progr. med. Genetics 9:* 113-58.

11   Genetics of Behaviour
[1] Tinbergen, N. (1972). Functional ethology and the human sciences. (Croonian Lecture for 1972). (Preprint).
[2] Davis, B. D., and Flaherty, P. (eds) (1975). *Proceedings of the Seminars on Human Diversity: Its Causes and Social Significance.* Amer. Acad. of Arts and Sciences, Boston, p. 84 ff.
[3] Ibid.

[4] Lorenz, K. (1966). *On Aggression* (English translation of *Das sogenannte Bose*). Methuen, London.

[5] Davis and Flaherty (eds), op. cit.

[6] Ibid., p. 91 ff.

[7] Gottesman, I. I., and Shields, J. (1972). *Schizophrenia and Genetics: A Twin Study Vantage Point*. Academic Press, New York.

[8] Heston, L. L. (1970). The genetics of schizophrenic and schizoid disease. *Science* *167*: 249-56.

12   Genetics of Power

[1] Darlington, C. D. (1964). *Genetics and Man*. Allen and Unwin, London.

13   Human Diversity

[1] Yoshida, A. (1973). Hemolytic anemia and G6PD deficiency. *Science 179*: 532-7.

[2] Gabor, D. (1972). *The Mature Society*. Secker and Warburg, London.

[3] Eysenck, H. J. (1975). *The Inequality of Man*. Fontana/Collins, London.

14   Of Good and Evil

[1] Wilson, E. O. (1975). *Sociobiology: The New Synthesis*. Harvard University Press, Cambridge, Massachusetts.

15   Vision of a Million Years

[1] Darwin, C. G. (1952). *The Next Million Years*. Rupert Hart-Davis, London.

[2] Dean, G. (1963). *The Porphyrias: A Story of Inheritance and Environment*. Lippincott Co., Philadelphia; Pitman Medical Publishing Co., Great Britain.

[3] Mc Devitt, H. O., and Bodmer, W. F. (1974). HL-A, immune-response genes, and disease. *Lancet 1*: 1269-75.

Index